the DAIRY GOAT handbook

FOR BACKYARD, HOMESTEAD, AND SMALL FARM

ann starbard

Voyageur
Press

Dedication:

To Pippy and Simon

Quarto is the authority on a wide range of topics.

Quarto educates, entertains and enriches the lives of our readers—enthusiasts and lovers of hands-on living.

www.quartoknows.com

First published in 2014 by Voyageur Press, an imprint of Quarto Publishing Group USA Inc., 400 First Avenue North, Suite 400, Minneapolis, MN 55401 USA
Telephone: (612) 344-8100 Fax: (612) 344-8692

quartoknows.com
Visit our blogs at quartoknows.com

Voyageur Press titles are also available at discounts in bulk quantity for industrial or sales-promotional use. For details write to Special Sales Manager at Quarto Publishing Group USA Inc., 400 First Avenue North, Suite 400, Minneapolis, MN 55401 USA.

ISBN: 978-0-7603-4731-7

Library of Congress Control Number: 2015932391

Acquisitions Editor: Elizabeth Noll
Project Manager: Jordan Wiklund
Design Manager: James Kegley
Cover Designer: Kim Winscher
Layout Designer: Kazuko Collins

On the front cover: A mother and her kid graze in the grass.
On the back cover: Newborn kid ready to play (top); chevre (bottom left); does enjoy the whey from cheese-making (bottom right).
On the frontis: Goats mug for the camera.
On the title page: Some goats have an almost preternatural sense to understand your needs or language. Talk to them often!

Printed in China

10 9 8 7 6 5 4 3

Contents

Introduction

Goats are truly one of the most adaptable and productive domesticated animals on our planet. They are found thriving in cold, mountainous areas; on warm, tropical islands; and so many places in between. They thrive on landscapes and feed on plants that challenge other animals. Domesticated thousands of years ago, goats continue to find their way into the spotlight. Goats are said to be one of the most popular small farm animals in the United States today. The reasons are many.

A few dairy goats in the backyard can supply a family with nutritious and flavorful milk to drink and make into other foods. They can also provide meat from the raising and slaughtering of offspring, and manure for fertilizing gardens. Goats will eat or browse overgrown and invasive plants, cleaning up underutilized species while feeding their bodies rich nutrients. At the same time, with their smaller hoof print, agility, and selective feeding habits, goats, in limited-resource areas with sparse vegetation, can still provide for us humans. Thus the dairy goat's

Dairy goat eating mineral-rich nettles (*Urtica dioica*).

Goats have many endearing qualities.

Raising dairy goats is rewarding, challenging, and fun.

efficient and adaptable nature is perhaps part of a local answer for the environmental, economic, and food safety challenges we face today.

Of course, some feel goats have a downside: that they "question authority." Many are familiar with the biblical references to sheep following on the right, while goats swing and jump to the left, listening to their own tune. This unpredictable and devilish nature of goats is well expressed in history, literature, and even pop culture. The frolicking forest god of Greek mythology, Pan, is a classic goat-like figure, while a favorite 2013 Doritos commercial features a manically munching goat. Story lines change, yet the impish goat behavior continues to trend. Why?

Goats have a curiosity that is often misunderstood. Take some time to be with goats and you will soon discover their endearing characteristics. Innocently, they create a unique and lasting bond. They are fun companions despite their independent spirit. Bring goats into your life and you will quickly learn many lessons both about goats and yourself. In addition to honing your basic animal-raising abilities, goats will challenge your wit, patience, and fence-building skills. You will develop GOAT-titude! GOAT-titude is the mental state that is necessary to thrive with the inquisitive, adventurous, adaptable, impish, intelligent, laughable, and energetic behavior of goats.

The Dairy Goat Handbook will help you learn about dairy goats and how to add their lively spirits, nutritional food products, and resourcefulness to your life.

Let's look at the advantages of dairy goats. First and foremost, they are adaptable and hardy animals. They thrive and reproduce with simple but attentive care. Their smaller size at maturity makes goats easier to handle than other food-producing animals, and they are generally gentle. Their space and feeding needs are minimal. Overall, goats cost less money to acquire and keep.

They eat a variety of plant material: grasses, weeds, leaves, small browse, trees, and bushes. They are able to make use of less expensive feeds, which surprisingly are often higher in mineral content than conventional feeds. Goats are selective feeders, making them good brush-clearing animals. They eat a little here and a little there, acquiring a variety of plants to meet their dietary needs.

The milk they produce is nutritious and flavorful. A single goat of good breeding can easily produce enough milk for a family to drink and use for making dairy products such as cheese and kefir. Goat milk is more easily digested and does not produce as many allergic reactions as cow milk. Production and body type can be improved in a relatively short period of time.

Develop your GOAT-titude!
Danielle Mulcahy

Lastly, goats are good family animals. They teach responsibility and decision making. Everyone learns that an animal getting good food and proper care will thrive and make delicious, healthy food for the family to eat. Children learn the importance of independent yet cooperative work, knowing they play an important role in taking care of a live, productive animal. The entire family takes part in creating a successful and enjoyable enterprise.

Life is certainly about balance, so let's look at some of the disadvantages of dairy goats.

They are a commitment. Generally, a goat needs to be milked twice every day for 10-plus months of the year. The life expectancy of a goat is ten to fifteen years.

Goats need to be properly managed to avoid overgrazing and denuding plant growth. Without good fences or proper attention, roaming goats can cause damage to gardens, yard plants, and even cars! Goats can be spoiled and develop bad attitudes, which may lead to behavioral problems. Poor or insufficient feeding and care can also cause difficulties.

Dairy goats normally reproduce every year, leading to overpopulation if offspring are not sold. Goat-savvy veterinarians and medications approved for milking goats can be difficult to find. Quite frankly, there will be bad days and frustrating moments. But these negatives can easily be minimized with knowledge, work, and planning.

This book is designed to teach both beginner and advanced goat keepers about dairy goats. There is basic information for getting started and raising goats, with advanced sections and more details on specific subjects, such as management, breeding, health care, and milk production. Each chapter could easily be its own book. I highly recommend that everyone find an animal mentor—someone raising goats—as experience is generally the best teacher. Find an agricultural club in your area or at least an Internet forum to help answer detailed questions and discuss situations. Keep a barn journal. Develop and strengthen your common sense. Goats require shelter, food, companionship, and some nurturing—basically the needs we all share. Provide this care, pay attention to your animals' needs, and you will receive plenty of benefits from your dairy goats.

Yes, there is daily work and a commitment to raising dairy goats. This work, though, is wonderfully rewarding, positively challenging, and just plain fun! Be curious, observe, act, and enjoy—in other words, add GOAT-titude to your life!

Chapter 1

Getting Started

"Well begun is half done." How often have we heard this? The decision to get dairy goats or advance in your goat keeping requires some discussion and thought. Where do you begin? Approach the process using the following considerations, adding components unique to your situation. Be realistic and practical. Everyone who will join in the animal care should be involved in the decision making. Please be honest, communicate, and don't make assumptions. Goats should not suffer due to our follies. Be clear, and understand your decision. Enthusiasm can be contagious—share the workload and commitment, and everyone will benefit.

Visit goat farms, talk with people keeping goats, attend agricultural fairs and goat shows, and find a goat mentor. Learn about the realities of your ambition while you explore the possibilities. Be considerate of people's time, and offer to compensate them for their expertise. Showing respect and appreciation for an experienced farmer's time and knowledge goes a long way toward creating an invaluable long-term relationship.

Considerations

Set a goal. Setting a goal isn't for everyone. That's fine. Simply having a couple of goats to be companions, provide milk, or browse a vegetated area is a good way to begin. FYI: These are all goals, whether you like the word or not. Goals are fluid, moving and living. Setting and achieving goals with your goats are a good way to teach children the merits of balancing work and accomplishment with fun

Find a goat mentor who will teach you the realities of your ambition.

and play. Strategize with all family members on setting your goat goal(s). Keep in mind you need to have at least two goats, as a goat is a social animal that needs a companion. Write your goals on a white/chalkboard in the goat house. Come up with new goals and challenges, making sure they are practical and within your resources. Most importantly, be realistic and have fun! We'll look more specifically at getting animals to meet your goals in Chapter 2, Breeds and Selection.

Determine your available time. Caring for animals takes time. Mature dairy goats require milking. Kids can nurse off their moms for a bit of time, but eventually you will have to milk. Milking is generally done twice a day, about every 12 hours. You can cut down to once a day with a kid nursing part time, or with lighter milk production in late lactation. In addition to milking, you'll need time to acquire and distribute feed; set, move, and fix fences; clean and bed the pen; and perform routine health maintenance, such as hoof trimming, parasite monitoring, and vaccinations. Will you choose to make cheese, soap, yogurt, and other products from the milk? You'll need to figure these jobs into your day. Start with the goat raising and milking. Develop a routine. Phase in additional tasks as you become efficient and acclimated to your animals. Please just be aware that you will need to make a time commitment with dairy goats; don't shortchange the animals, yourself and family/friends, or the end product: the milk. Dairy goats are very agreeable; work with others to share the workload.

Determine your financial means. Getting started in dairy goats at any level generally requires financial input. Yes, there are many ways to save money, including doing things yourself and bartering, but you will need to acquire healthy animals, secure feed and water, build an adequate shelter, and fence off an exercise or pasture area. Generally the old adage "you get what you pay for" does apply here. Research your options, consult with your goat mentor or veterinarian, and make informed decisions. Include a cushion for your budget expenses, as construction or feed costs can rise when you're midstream. We'll look more at the real numbers in Chapter 5, Business.

Make a commitment. Argh! That scary word! It's okay. That's why you are reading this book. As with most things we fear, learning about and sharing the task is the key to finding the joy and pleasure of the commitment to dairy goats. The benefits outweigh the costs in so many ways. Animal care is calming and almost therapeutic when you make the commitment of time and energy. Milking brings a deliberate slowdown to the day, a time to be with your goats and family/friends that are sharing the work. Plus you will have delicious, nutritious milk to drink and use to make other dairy products. Have control of the situation with a proper goat house, plan for feed and water, functioning and practical fences, and "organized labor" to make the commitment interesting and pleasurable to helpers of all ages.

Determine your available space. This plays a big role in determining your goals. There are many ways to keep dairy goats on minimal acreage. Ideally, goats like to move and forage, but they can thrive in a small pasture system where the majority of their feed is brought to them. Determine how to best utilize your available space to keep healthy dairy goats and achieve your goals. Keep in mind that goats generate manure. Have space for a compost area, find friends to take the manure, or use this rich resource for income generation.

Determine whether goats fit into your current family of animals. Goats are good companion animals, but every animal has its individual personality and needs. Decide if your current animals have the attitudes to accept goats into their space. There are advantages and disadvantages to mixing farm animals. We'll look at interspecies mixing in Chapter 4, Management.

Determine who has DIY skills. Are you or someone in your circle talented at carpentry, building, or fixing things? If not, find someone you can hire to do this type of work. There are several tried-and-true "goat proof" structures, but goats challenge even the best of carpenters. Frustration can rise at their cleverness and tenacity. These critters are just so curious and inquisitive. Knowing how to add another board or mend a fence is important. Have some basic tools and knowledge to at least temporarily fix something until a trained carpenter can arrive.

After looking at these considerations, go back to your goals and see if they are realistic and attainable. Can you strive to meet the challenges? If not, reassess your goal, knowing your available resources. You may have to break the goal into steps, keeping the big picture in mind. I believe in giving your absolute best to Plan A. This is the way we grow, learn, and achieve. Yes, that power of a positive attitude. There is, though, merit to having a Plan B lurking in the background. Be flexible. Opportunities *and* road blocks pop up when we least expect them. Go back to these considerations and reform goals as needed.

If you are a beginner, spend time with goats to make sure you don't have allergies to goat dander, hay, and other allergens inherent in a goat environment.

Okay, you have your goal; now what? Decide on housing, pen layout, feed sources, fencing, and lastly, where you will acquire your goats. You can work on these elements simultaneously. Be prepared by having a pen, proper fence, feed, and supplies ready before you bring goats home.

Housing

Goats need shelter from the elements. The most critical needs for goat housing are good ventilation and protection from wetness and drafts. Some farm animals, such as cattle, sheep, and horses, fair okay standing in the rain and snow. Dairy goats, simply lacking the insulating and water shedding abilities of other animals, do not. They bolt for shelter with the first drop of rain. Perhaps they are really wicked witches from the west and will melt if they get wet!

A three-sided structure is adequate in warmer climates.

Newborn and young goat kids are particularly sensitive to drafts.

Follow your considerations from above when planning your goat house, remembering that you will be milking and tending your dairy goats at least twice a day. Build a goat house that is functional, pleasant, and welcoming. Size your building for the maximum number of animals in your plan, with 16 square feet per animal as a minimum. Include ample windows for sunlight and airflow.

Goat shelters can be quite simple for a few goats in your backyard. A three-sided building in warmer climates is adequate.

During a cold winter or bad weather, more protection, such as a door or a securely hung heavy blanket, is needed to keep cold wind and snow from chilling the animals. Plan an overhang on one side to place a folding milk stand if you are tight on space. Feed can be stored in a garage or other outbuilding. Use tightly covered containers to keep unwanted critters out. You may fancy a finer home for your caprines and have the skills and/or finances to achieve this. Regardless, build a structure that is healthy for your animals and pleasing to you.

Dairy goats kept in suburban settings are becoming more popular. Learn the zoning regulations in your area. Goats can make good neighbors, with proper facilities and management. Keep an appropriate number of animals, build a goat house that blends into the established surroundings, control pests, and care for the goats (both milking and feeding) in a timely manner to minimize noise. Close neighbors and breeding bucks don't usually mix well, so plan to take your goats elsewhere for breeding.

Greenhouse barns cost less initially to build than traditional wooden frame structures. These buildings retain heat in the winter, making a good area for birthing and growing kids. Proper airflow and ventilation are important, especially in the warmer seasons. Plans for larger herds will need to consider inclusive areas for feed storage and feed trough/racks, easy access to water, smaller pens or jugs for kidding and health care, ease

Some prefer a fancier goat house.

of cleaning and bedding storage, flexible pens to house younger goats as they grow, a separate area for milking, and a possible creamery or production area.

Again, look at your goals, considerations, and future plans. Commercial licensed facilities will have to meet federal, state, and local guidelines and regulations, which are updated and changed frequently. Check with local health and agricultural officials to learn about the current requirements for your area. Consult before you construct so you don't waste time and money replacing or rebuilding if you plan to sell your goat dairy products commercially.

This suburban goat house blends into a small space.

This open pen in a warm climate has predator protection.

The open housing area here has keyhole feeders on the pen perimeter.

Pen Layout

Create an easy flow for animal movement with a good pen design. Dairy goats are most productive and healthy when they are kept calm and moved with ease. Author Temple Grandin has brought animal behavior to the forefront with clever ideas to ease the stress of animals. In her book, *Animals Make Us Human*, she states, "Everyone that is responsible for animals . . . needs a set of simple, reliable guidelines for creating good mental welfare that can be applied to any animal in any situation, and the best guidelines we have are the core emotion systems in the brain. The rule is simple: Don't stimulate rage, fear, and panic if you can help it, and do stimulate seeking and also play. Provide environments that will keep the animal occupied and prevent the development of stereotypes."

Keep these thoughts in mind as you design your pen, alleviating animal stress as you move the goats for milking and feeding. Create areas for exercise and positive stimulation. Develop a pattern or routine for daily chores. Dairy goats who feel safe, secure, and serene will produce the best milk. You want happy goats!

Provide benches and areas for the goats to climb and rest. Be creative, making sure to build sound and secure structures. Benches are ideal as young goats can hide underneath while the adults lie or stand on them.

Hay feeders and watering areas need to be at an accessible height. This varies for different breeds. Keyhole and slant-board feeders minimize hay loss. Goats like to stand on a ledge to reach their feed. Automatic waterers have an initial expense, but I think they are priceless if you are keeping ten or more animals. Carrying water or thawing frozen buckets gets old really fast, especially as the goat keeper ages. Look for automatic waterers that have a heating element option if you live in colder regions. You will *not* regret your decision to spend extra money on waterers!

Goats need salt and minerals. There are several designs available for having either free choice or block salt and minerals accessible to your goats. Salt and minerals need to be sheltered from precipitation. As with feeders and waterers, keep salt and minerals at a proper height for your goats. Be sure to put small salt bricks in your kid and buck pens.

This small pen is appropriate for dwarf breeds.

Gates can be used to temporarily create a smaller pen, such as a group area for before or after milking or when separating the herd to hoof trim, vaccinate, or check for parasites.

The number of mature does you have determines the layout of the milking area. One stand for every four to six milking goats is a good ratio. Will you be hand milking or using a machine? Running and hot water are excellent considerations for cleaning milking stations and equipment. Legal sale of fluid milk and processed dairy products require strict and detailed setup of milking areas. Consult with local, state, and, if necessary, federal officials in the initial stages of planning a licensed milking facility. Please see the resource page (180) for more information.

Climb over gates at the hinges to decrease the tendency of gates to sag.

Simple wooden benches are a beneficial addition to a goat house. *Danielle Mulcahy*

Feed Storage

Feeding will most likely be your largest daily expense, especially if you have limited space for goat browsing. Buying in bulk or partnering with other animal keepers will help reduce your bottom line, as long as there is local availability and you have room for feed storage. Consider how you can store feed so you can take advantage of lower prices and obtaining the highest quality possible. Quality feeds will make the difference between good milk production, in terms of both quantity and quality. Finding good quality feeds for the best price can be a challenge. Research cheaper alternative, locally available feeds, and how you can store these for use during the non-growing season. If you are simply housing a few goats in a small shelter, feed can be stored in a garage or other outbuilding. Use tightly covered containers to keep unwanted critters out of grain and seeds. Be sure to rotate feedstuffs and discard any grain or hay that becomes contaminated, moldy, or wet. We'll discuss actual feedstuffs and nutrition in Chapter 3, Feeding.

A greenhouse barn is warm in the winter, and the open pen makes for easy feeding with round hay bales.

The right goat pen depends upon the size of your herd, the breed of the goats, and any particular features you want to include. As long as the goats can't get out, though, it'll probably be a success! *Shutterstock*

An automatic waterer is invaluable for larger herds.

Keep mineral feeders properly adjusted for the height of your goats.

Hanging a salt block provides nutritional supplement and amusement.

Gates ease handling of animals and break a larger pen into smaller areas.

This is a grain bin for bulk delivery of grain.

Fencing

Many goat keepers consider fencing the most difficult part of keeping goats, well, *in*! The options for fencing are much better today than in the past. Technology and advances in fencing materials offer cost-effective options that, when correctly installed and used, work very well to keep goats in their proper place. I train our young goat kids to an electric net fence when they are 2 months old. Read more about this in Chapter 8, Kid Care. The key here is training the goats when they are young. Our herd of seventy adult goats lives behind two strands of simple electric fence, which they rarely challenge.

Ill-behaved (a.k.a. spoiled) goats that were allowed or encouraged to jump fences when younger will simply not respect any fence. Proper discipline and training at the earliest age makes a huge difference. Consider the advantages and disadvantages of various fences. Visit farms and fence dealers to see what options may work for your situation. Keep in mind that all fences need to be properly

Double-stranded electric fence that is semi-movable works well to contain trained goats.

Electric fences need to be properly maintained and cleared of plants.

Fixed-knot, high-tensile fencing with top electric wire is a permanent solution.

installed and maintained. Weeds growing on electric fences diminish the available shocking voltage, as rain and morning dew will steal current flow.

Electric fences, including the net fences, do not work as well in very dry climates. In this case, soil moisture is insufficient to complete the current flow for an animal to get shocked. Most electric fences need a minimum of 3,000 volts. Check with the manufacturer for specific details. Here are some fence choices that work well with goats.

Permanent: High tensile. Expensive, permanent; can be used as a perimeter fence.

Semi-permanent: Cattle panels attached to T-posts. Secure cattle panels to T-posts placed every 8 feet. This fencing is great for younger kids. Posts are on the outside of the fence. A single line of electric fence can be strung at 8 to 18 inches off the ground and offset 4 to 6 inches. This hot wire works particularly well with older goats that like to rub or "climb" the fence.

Detail of fixed-knot fencing.

Semi-movable: Two strands of electric wire attached to T-posts and fiberglass poles. Set up as an exercise area or perimeter fence. The fence will need to be mowed and maintained. See photo on page 20.

Movable: Net fences. These have to be moved frequently and require good maintenance of the fence and electrical system. Be aware that goats can get stuck in them, panic, and be injured, especially if they are not trained to this type of fence.

Solar chargers can be used to electrify fences away from a building. Research the appropriate size charger and electrical output necessary for the fence you are installing. Be sure to test the attached fence for adequate voltage.

Electric wire fencing is an effective way to keep your goats on your property.

Cattle panels attached to T-posts make a sturdy fence, particularly for younger animals.

Finally, Getting Your GOATS!

You have put a lot of energy and time into making the decision to get goats with setting a goal, planning their care, housing, pen setup, feed, and fencing. Now comes the fun part: getting your goats! Three criteria should be considered:

1. Buy animals that meet your goal.
2. Obtain healthy, strong goats from a reputable person.
3. Pay what you can afford.

Will you select registered purebred goats? Can mixed breeding fit your needs? Will you buy locally, or do you want animals from another area? The next chapter discusses dairy goat breeds and how to select animals.

The 5 Ps: Prior Planning Prevents Poor Performance

Left: A movable fence, such as this net fence, is useful for making smaller, rotational browse/pasture areas.

Below: Solar fence chargers are useful for fencing browse areas that are a distance from line electricity.

Chapter 2

Breeds and Selection

There are currently eight primary dairy goat breeds in the United States: Alpine, LaMancha, Nigerian Dwarf, Nubian, Oberhasli, Saanen, Sable, and Toggenburg. Everyone has a favorite breed for a variety of reasons—personality, appearance, thriftiness, milk qualities, size, or local availability. I invite you to spend some time with each of the breeds if you can. Talk with goat owners and do your own research, even if you are looking for crossbred animals. Do characteristics of a certain breed "talk" to you? Keep an open mind and consider all the possibilities. As with your goals, be flexible.

You may decide you like the regal look of Oberhaslis, and then years later, find a Nubian that you just connect with. Seize the opportunities as they come. I started with Saanens and Alpines and was quite happy with these breeds. A local woman was moving and had to find a home for her eight LaManchas. I added these quirky, short-eared goats to our herd and quickly fell in love with them.

The following short descriptions and photographs will give you some basic information on the breeds.

ALPINE

This dairy breed is medium to large size, with mature does weighing at least 135 pounds

Alpines can be attentive and even bossy in their moods and wants—watch them carefully!

and measuring a minimum of 32 inches at the top of shoulders (withers). Alpines are an independent breed, gracefully aggressive and passively attentive. They produce an average of 2,400 pounds of milk a year with a 3.5 percent milk fat and 2.9 percent protein. They have erect ears, short to medium hair, and range in many color combinations.

Alpine Color Descriptions

Cou Blanc (coo blanc) – French for "white neck." White front quarters and black hindquarters; black or gray markings on the head.

Cou Clair (coo clair) – French for "clear neck." Front quarters are tan or off-white with **black** hindquarter.

Cou Noir (coo nwah) – French for "black neck." Black front quarters with white hindquarter.

Sundgau (sundgow) – Black with white markings on face, underbelly, or legs.

Pied - Spotted or mottled.

Chamoisee (shamwahzay) – Brown or bay body with characteristic black markings, such as back stripe, feet, legs, and tail. Spelling for male is chamois.

Two-Tone Chamoisee – Light (white, tan, off-white) front quarters with brown or gray hindquarters.

Broken – Any variation in the above patterns stippled with white areas is described as a broken, such as a broken chamoisee.

LAMANCHA

This breed is known for its short, or basically nonexistent, external ear. LaManchas were developed in the western United States in the 1920s by crossing short-eared Spanish bloodlines with various dairy breeds to produce a good

LaManchas make up for in attitude what they lack in, well, ears. They are adaptable, friendly, and resourceful dairy producers.

dairy goat. The result, in addition to productivity, is an adaptable, friendly, and resourceful goat. The LaMancha ear can be described as "Gopher" or "Elf." A gopher ear is about 1 inch in length with little or no cartilage, and the only acceptable form of ear for a registered buck. The elf ear is about 2 inches in length and can have some cartilage shaping the ear. This breed has short to medium fine hair, with all and any colors and combinations. They average 120 pounds for a mature doe with a height at the top of the shoulder of 28 to 30 inches. LaManchas are a sturdy breed, with admirable curiosity and personality. The LaManchas in our herd love to be out in the fields browsing. They often lead the Alpines and Saanens into the pasture and stay out eating long after the others have lain down under a shade tree. Milk production averages 2,200 pounds annually with a 3.7 percent milk fat and 3.1 percent protein.

NIGERIAN DWARF

This smaller breed is gaining in popularity due to its ease of handling, adaptability, and rich milk. The ideal weight of a mature doe is about 75 pounds, and the average height is 17 to 19 inches. Nigerian Dwarfs have bodies with balanced proportion, similar to the larger dairy breeds. They have upright ears and short to medium hair. The coat colors vary with all and any color combination and patterns, with the main colors being black, chocolate, and gold. They are prolific breeders, oftentimes giving birth to three or four kids. The milk quantity averages 600 pounds a year, with an average milk fat of 6.3 percent and protein of 4.3 percent This breed is ideal for goat keepers of all ages, and especially for homestead keepers making dairy products from the milk. Nigerian Dwarfs mix well with other goats and adapt well to many settings, including urban environments.

Nigerian Dwarf goats are small, but their popularity is growing due to their feisty and fun-loving natures, and the fine, sweet milk they produce. *Ellen Gould*

Nubians produce rich milk and have strong personalities. They are the most popular dairy breed in the United States.

NUBIAN

Nubians are currently the most popular dairy goat breed in the United States. The breed has a long, interesting history. The Anglo-Nubian is of British breeding, crossing Old English goats with lop-eared breeds from India and North and East Africa. The result is this medium- to large-framed goat with characteristic long, pendulous ears and a convex Roman nose. The mature doe should weigh at least 135 pounds and stand an average of 30 to 32 inches at the withers. They are a hardy and proud breed, with strong personalities. They can also be quite vocal. The hair coat is short and of any or all colors, solid and patterned. The richness of the Nubian milk is one reason for the popularity of this breed, especially for making cheese and milk products. Nubian milk averages 4.5 percent milk fat and 3.7 percent protein with an annual production of 1,980 pounds.

OBERHASLI

Oberhasli goats are a medium size, averaging 130 pounds and standing 28 to 30 inches at the withers for a mature doe. They have erect ears and a short to medium hair coat. Oberhasli goats have a striking body color: solid bay (deepest red is most desired) with black markings on their face, back, dorsal stripe, belly, udder, lower legs, and tail. Females may be black, but bucks must be the traditional bay color. Oberhaslis, as a breed, are vigorous and lively, yet cooperative. They mix well with other breeds. They average 2,200 pounds in milk production, with 3.6 percent milk fat and 2.9 percent protein.

Oberhaslis make fine companions to other domestic goat breeds. *Tricia Smith*

Pale and regal, Saanens may often dominate a herd with their stubborn, energetic personalities.

Like their Swiss forefathers, Toggenburgs are calm and friendly—neither overly dominant, nor unusually submissive. You might call them the neutral citizens of the herd!
Jay Iversen

SAANEN

Saanens are a productive dairy goat with a medium to large frame. The mature doe weighs around 150 pounds and stands 32 to 34 inches at the top of the shoulders. They have erect ears and distinctly graceful beards, particularly as they age. Saanens are solid white to cream with occasional black spots on the nose, ears, and udder. They are a vigorous, strong-willed breed, often dominating the herd. This is especially seen at the hay rack. The annual production average for Saanens is 2,600 pounds of milk with 3.3 percent milk fat and 2.8 percent protein.

SABLE

The Sable breed is a variant of the Saanens, being recently (2005) recognized by the American Dairy Goat Association (ADGA). Sables are basically Saanens with a gene variant for coat color. "The Sable coat and markings may vary from solid tan to darker colors, patterns and combinations," according to Sable Breed standards. The minimum weight is 135 pounds and height is 30 to 32 inches for a mature doe. The production data is similar to that of Saanens.

TOGGENBURG

Purported to be the oldest-known dairy breed of goats, this breed originated in the eastern Swiss valley, Toggenburg. These goats are calm and friendly, with warm, kind eyes. They are a strong, healthy breed of medium size, with erect ears. Mature does weigh at least 120 pounds and are 28 to 30 inches at the withers. The coat of brown (fawn to dark chocolate) is highlighted with characteristic white markings on the face, ears, muzzle, legs, and tail. Toggenburgs historically have medium to long hair, though shorter hair is being favored recently. They average 2,260 pounds of milk, with 3.2 percent milk fat and 2.7 percent protein.

There are interesting stories to go with each of the breeds and how they became what they are today. Each breed has a national club whose information is given in the resource section of this book (page 180).

Once you're ready to start looking for animals, read through the following guidelines. These pointers will help you through the process. Starting with strong, *healthy* animals is important. You have put time into making decisions and goals, setting up a goat house, and teaming up with family and friends to help share the work and milk! Look at the Goals for Selecting Dairy Goats Chart. Write in your own key points, and take this with you when looking at animals. Start with what you can afford. Improving *your* herd is part of the fun of keeping dairy goats.

WHERE TO LOOK

First, local fairs will often have a goat show or at least goats on display. Second, contact local organizations such as 4-H, Grange, or County Extension. Third, attend area farmers markets. Many commercial goat farmers attend local markets to sell dairy products. Also, check postings at local agricultural stores or on the Internet for local goat forums. Warning: Be extra careful when buying off an Internet list. While there are many fine, healthy animals being sold online, be aware that many unhealthy animals are sold also. Do your homework and politely walk away if you question the animals or the seller. I do not recommend buying animals from an auction to start a dairy herd as there are many potential problems from buying from an unknown source. Auction buying has a purpose; starting or increasing a dairy goat herd isn't one of them.

Look at your goals and considerations, decide the breed or crossbreeds that you want to raise, think about the milk production and qualities you want, and go have fun! Visual selection of dairy goats requires keen observation, in-depth questions, and some experience. Beginner goat owners may ask an experienced person to come with them to look at animals. Learn about ideal dairy character by attending local fairs and observing the dairy goat and cow shows. Listen to the judge's reasoning for placing animals in the ring. Ask questions once the animals are out of the show ring and the handler is finished showing for the day. Most animal owners are interested in sharing knowledge of their animals, particularly if they have animals or offspring for sale. Be polite, considerate, and understand the show schedule.

Breed Average Production

Annual Average	Milk (pounds)	Milk Fat (%)	Protein (%)
Alpine	2,500	3.4	2.9
LaMancha	2,200	3.7	3.1
Nigerian Dwarf	600	6.3	4.3
Nubian	1,980	4.5	3.7
Oberhasli	2,200	3.6	2.9
Saanen	2,600	3.3	2.8
Sable	2,200	3.5	2.8
Toggenburg	2,260	3.2	2.7

Goals for Selecting Dairy Goats

Goal	Animal Type[1]	$$	Where to Look[2]	Desired Age	Notes[3]
Homestead use	PB, CB	High–Low	B, CF, H	Any	Good personality
Expand commercially	PB, CB	High–Medium	B, CF	Any, 2- to 3-year-old is preferred	Fit into larger herd
Show, sell breeding stock	PB	High	B, CF	Any, 1- to 3-month-old is preferred	Excellent genetics, look at family lines
Family fun goat	PB, CB	Medium–Low	CF, H	Any	Easy personality
Improve herd	PB	High–Medium	B, CF	Any	Sound genetics, look at family lines Socially adjusted
Companion animal	PB, CB	Medium–Low	CF, H	Any	personality
YOUR GOAL:					

PB- Purebred, CB- Crossbred, B- Breeders, CF- Commercial Farm,
H- Hobbyist. Always buy a healthy sound animal.

Parts of a goat's body

chine · loin · hook (hipbone) · withers · rump · face · muzzle · rib · pinbone · jaws · point of shoulder · point of elbow · flank · barrel · rear udder · heart girth · fore udder · milk vein · teats · hock · knee · dewclaw · toe · pastern · heel

Learn the correct names of your goats' bodies to share information with veterinarians and goat farmers.
Courtesy Barbara Carter

PREPARING TO VIEW GOATS

Wear clean clothes and clean shoes/boots when traveling to look at animals. The owner may ask you to wear plastic boots over your shoes to stop the spread of diseases. These boots can be cheaply purchased online or at agricultural stores. Buy a box and keep the boots with you to wear on farm visits.

Let's learn some simple goat anatomy.

Goat Anatomy

1. **Teeth** Goats have a total of thirty-two teeth: eight incisors on the lower jaw only, twelve premolars, and twelve molars (six of each on the upper and lower jaws). The upper front jaw has a dental pad (firm soft tissue) that helps the incisors to tear forage. The young incisors of goat kids are replaced by adult incisors as the animal ages. You can use the teeth to roughly age a goat until it is four years old. The molars are used to grind the feed during cud chewing. The teeth of older goats may develop sharp points that can cause problems and may need to be filed down.

2. **Wattles** Wattles are small, vestigial skin appendages, generally found on the neck. Vestigial means they have lost their function through the course of evolution. Some believe wattles were part of the salivary gland system; others say they evolved from fish gill slits. Regardless, their only current function seems to be appearing on lists of frequently asked questions about goats.

3. **Knees and Hocks** The goat knee is similar to our wrist, the hock to our heel. These joints are comprised of many small bones puzzled together. The leg bones from these points to the feet are two metacarpal bones (front) and two metatarsal bones (rear) fused together. The foot structure is similar to bones of the third and fourth digits of the human hand/foot.

4. **Withers** This is the ridge on the back between the shoulder blades and used for height measurements. This is sometimes referred to as top of the shoulders.

5. **Heart Girth** This is the measurement around the fore-barrel taken right behind the front legs and the shoulders. This measurement is used to estimate the goat's weight.

6. **Rumen** Goats are ruminants, meaning they have a multi-chambered gastric (stomach) system. The rumen is the first chamber that feed enters after chewing and swallowing. The rumen is basically a big fermentation vat, with enzyme-producing microorganisms. The enzymes break down plant fibers, making nutrients available to the goat's digestive system. The rumen begins functioning when goats are around 3 weeks of age.

7. **Hooks** This is the term used for the hip bones, which are located on either side at the end of the rack of vertebrae.

8. **Pin Bones** These are located on either side of the tail head.

9. **Rump** The pelvic region between the hooks and pin bones.

10. **Udder** This is the goat's milk-producing gland, and it is split in half. Each half is a separate chamber, so milk does not flow from one to the other. Milk is secreted through an intricate cellular system, collecting in a milk cistern at the base of each half. Milk is retained in these cisterns through a series of tight muscular rings. The cistern is connected to a teat at the bottom of the udder, through which the milk is removed.

Older goats may require their teeth to be filed down if they become too sharp. *Shutterstock*

Wattles are vestigial appendages of unknown use and origin. *Shutterstock*

Selecting Animals

Younger animals may cost less initially, but consider the mortality rate as well as the expense, in terms of both money and time spent, of raising animals to a productive age. Some physical aspects of a dairy goat, such as udder shape and body conformation, are difficult to assess at a young age.

Before you start looking at a goat, ask the breeder about general herd health. Not everyone will give you an honest answer, so look for yourself as well. We'll talk about specific health tests later. Discuss the traits and qualities you are looking for. You may already know a specific animal through your research. If not, most breeders can help point you toward just the right one. Before going to look at animals, determine how many you want to buy and how much money you want to spend. Ask to see milk production records and records of any offspring or parents.

You are narrowing your search, so look now at the overall appearance of the goat. Does the animal appear healthy and content? Is the hair coat shiny and smooth or rough and ragged? Be aware that spring shedding can cause a temporary roughness. If the general appearance is satisfactory, begin assessing more detailed features.

Starting with the face, look for proper placement of the eyes and jaws. Eyes should be evenly located on the face and alert. Look for under- or overshot jaws. Jaw irregularities can cause feeding problems as the animal ages. Carefully examine lips and nasal areas for symptoms of active soremouth infection. Soremouth is a highly contagious skin disease caused by a virus.

Spread the lips and check the teeth for unusual wear or broken or missing teeth. Observe any lumps or growths in the jaw, ear, and neck area. Goats can have benign granulomas or cysts in these areas, but be aware that caseous lymphadentitis (CL) swellings are often found in these same areas. Avoid buying animals with CL, which is a chronically infectious disease caused by the bacterium *Corynebacterium pseudotuberculosis.*

Aspects of good dairy character in goats include a sleek neck flowing into a strong shoulder, chest, and rib area. Watch for "winged" shoulders—weak, loose shoulders are not desired. Goats can be easily weakened by lung ailments, so having open ribs and a strong chest floor helps defend against and heal from respiratory ills. Open ribs and depth of body allow for healthy rumen expansion and activity. Having plenty of body capacity for eating and rumen processing are important for dairy goats with good production. Generally, goats with extra body fat do not make strong milking animals. Select animals that have a good body capacity and are lean through the ribs and back line. An ideal dairy goat has a smooth, level top line and carries the frame evenly.

Look for a slight downward angle from the hip to the pin bone. Older goats often develop a steep angle, which can make kidding difficult. Width of hip and pin bones is important. While not directly associated with milk production, a strong, wide pelvic region is characteristic of a good dairy goat. This aspect allows for fewer birthing difficulties and an increased productive life. Also, if you are managing dairy goats with a natural feeding system of browse and pasture, a strong pelvic region is important for gathering browse. Goats will use their back, hips, and rear legs to stretch and retrieve plant branches. Having a strong, flexible, and agile musculoskeletal system, especially in the hindquarters, is therefore important.

Consider your goals particularly when examining the udder. Good dairy goats should have a well-placed and attached udder. Ideal teat placement varies for different milking situations. Regardless, teats should be evenly placed at the base of the udder. Look carefully to ensure each half has a single teat with no supernumerary, double, or forked teats. The udder should be carried slightly forward, definitely attached behind the naval area. Look for a strong rear udder attachment that is wide and high and flows firmly into wide escutcheon.

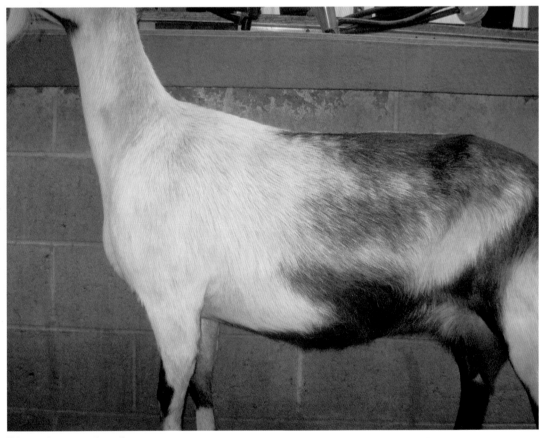

This goat has a smooth top line.

The legs and feet of a dairy goat need to be carefully examined. Even a highly productive goat will need to be culled if she can't walk. Look for straight front legs without any swelling at the knees. Hooves should be trimmed and even. Sturdy rear legs angle slightly above the knee and then straighten to a strong pastern area below the dewclaw. Avoid goats with over straight (posty) or bowed legs. Watch for weak pasterns and other poor physical features that may be hereditary.

Breeding bucks need to have an evenly hanging scrotal sack encasing two fully developed testicles. Bucks should have strong hips and rear legs.

Ask the owner if they or you can milk the goat if she is lactating. See how she reacts to someone touching the udder, and how the milk flows. This is particularly important if this is your first dairy goat or young children will be milking.

Other considerations for selecting dairy goats include determining the desired milk qualities. Individual animal milk statistics for daily production and fat and protein percentages need to be reviewed, as well as current and past history of somatic cell scores. Many commercial herds test milk monthly through a Dairy Herd Improvement program. Individuals can have samples tested periodically. The animal's production and health is determined both genetically and environmentally. Animals being fed high-quality alfalfa hay will not yield the same production on poor quality pasture. Be aware of the similarities and differences of the animals' current management and how you will manage them. Spend a few minutes assessing the animal's attitude when selecting a dairy goat. Will this animal fit into your goals and plans?

Once you have found animals you like, the price is right, and you are ready to seal the deal, ask again about health issues. Does the breeder have recent veterinary tests to show the health of the herd or the specific animals you are buying? Ask about parasite load and resistance, vaccination and deworming schedules, and any issues with soremouth, CL, Caprine Arthritis Encephalitis (CAE), Johnes disease, or Q fever. Tests are available for many of these diseases. Consider brucellosis and tuberculosis tests, and ask about any respiratory or diarrhea issues. If health test results are not available, the seller is not willing to test, and/or you have any doubts, ask an independent veterinarian to submit samples for you. Be sure to negotiate the purchase price with this cost in mind. There are veterinary diagnostic laboratories throughout the United States. The Washington State laboratory (waddl.vetmed.wsu.edu) is a helpful resource for testing goats.

Goats need strong and flexible hindquarters to stand and browse.

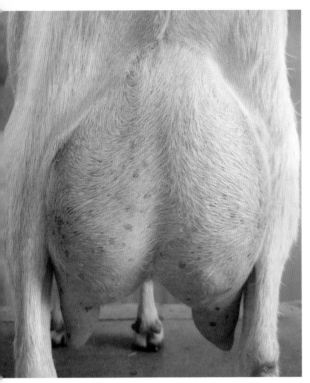

Here is an example of strong udder attachment with good structure.

Tips for Moving Animals

When you move lactating animals, expect milk production to drop in half initially. Change of environment, water, feed, and people are stressful. A goat may stop milking altogether if she is near the end of her lactation. To make the transition as easy as possible, ask the seller if you can purchase a bale of hay or some grains the animals are currently eating, be gentle with the goats, and have their housing or quarantine area at your property ready. Minimize stress from moving and resettling by refraining from transporting the goats in an open truck without some type of shelter from airflow or possible escape. Young goats can be put into large plastic dog crates, plywood can easily be attached to form a temporary wind block, or you can borrow a horse trailer on moving day. Be sure the transportation shelter is appropriate for your weather conditions, to protect the animals from precipitation and excessive airflow but allow ventilation to avoid overheating or suffocation. Bring shavings or straw for bedding. If traveling a distance, bring along water and hay, and stop along the way to allow the animals to safely drink, feed, and move around.

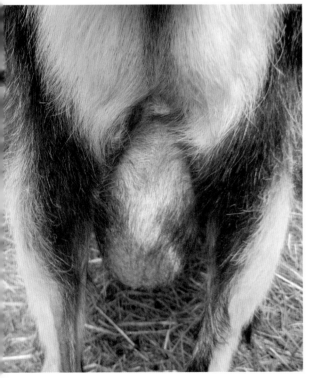

Evenly hanging scrotal sack with two fully developed testicles.

Traits to Avoid

Traits to Avoid	Reason
Overall weakness	Poor health, weak immune system
Rough hair coat	Possible parasites, internal and external
Irregular jaw	Difficulty tearing forage and cud chewing, especially with age
Broken teeth	Difficulty cud chewing, especially with age; uneven wear of teeth can produce mouth sores
Soremouth	Active, open sores; contagious
Caseous Lymphadentitis (CL)	Difficult to contain, spreads when abscesses burst; detrimental when selling breeding stock
Winged shoulder	Weak animal, poor body capacity, low milk production
Extra body fat	Low milk production, health issues
Sway back	Weak animal, poor body capacity, often leads to poor legs and feet
Steep rump angle	Often associated with weak udder attachments, difficult breeding and birthing
Weak udder attachments	Poor udder health, often leads to low milk production and problems with mastitis
Poor udder shape/ teat placement	Difficult to milk, problems with mastitis, low milk production
Swollen knees	Difficulty standing, walking, foraging, drinking water; prone to injury, general weakness and low milk production
Weak rear legs	Difficulty walking, foraging, jumping on milk stand; prone to injury, general weakness, and low milk production. For buck, poor breeding performance.
Posty or bowed legs	Difficulty walking and foraging; often seen with poor udder shape and attachments, low milk production
Poor feet, bad hooves	Difficulty standing, walking, foraging, drinking water; may lead to poor overall health
Bad attitude	Not any fun to work with!
Kicks when milked	No one will want to milk!

Chapter 3

Feeding

Feeding goats is important to all other aspects of raising them—breeding, birthing, growing, milk production, and health. With proper nutrition, dairy goats give birth to lively offspring, provide milk, and build a strong immune system. Goats are ruminants, but in many ways so different from other ruminants that we know—cows and sheep in particular. One major difference is that goats are browsers. They prefer to eat thick-stemmed forages such as leaves on trees, shrubs, and vines. They stand on their rear legs and reach for their feed.

They do enjoy pastures of mixed grass and legumes such as alfalfa, but they really eat through a field edge of roses, poison ivy, brambles, and other vines. For years, goat nutrition was simply a downsized cow ration or a sheep ration with slight variations. Happily, goats are now recognized as their own species.

Goats like to eat a variety of plants.

Many prominent research institutes in the United States are doing interesting nutritional work with goats, and learning and sharing with entities throughout the world.

Goats are classified as intermediate type ruminants in *Nutrient Requirements of Small Ruminants*, 2007. An intermediate ruminant "consumes mixed diets and forages based on season and opportunity. These animals prefer diets based on the concentrate components of plant material and will select shrubs and browse, but also possess the ability to digest cellulosic materials." As the term "intermediate" infers, goats are adaptable and have the ability to utilize a variety of feeds in diverse and challenging landscapes.

Gastrointestinal System

Let's look at the goat's gastrointestinal system. Being ruminants, they have a four-compartment stomach system consisting of the reticulum, rumen, omasum, and abomasum. Each of these compartments have specialized physical features and functions in the pursuit of converting plant material to nutrients the goat's body needs.

A goat uses her lips, teeth, and tongue to bring food into her mouth and swallow.

Goats have active salivary glands, aiding in the movement and digestion of food. Ingested food reaches the reticulum via the tubular esophagus. The reticulum has an interesting honeycomb-patterned lining. This compartment helps mix and move feed in many directions through its contracting actions. The reticulum shifts feed into and throughout the rumen, helps bring feed from the rumen to the mouth for cud chewing, and also moves feed further down the gastric system from the rumen to the omasum.

A doe uses her lips and tongue to bring feed into her mouth.

The largest of the four compartments, the rumen is a storage compartment and a big fermentation vat. A diverse population of active microorganisms breaks down the ingested plant material through secreted enzymes. A goat relies on active rumen flora to begin the digestion of plant material into life-sustaining nutrients. In many ways, feeding a goat is really feeding the rumen microorganisms. This is one reason why sudden feed changes cause problems. The rumen floras need time to adjust to digest new feedstuffs.

Contraction of the rumen and reticulum helps to mix and break down the material. The rumen contractions cycle rhythmically, one to two per minute. Slugs or boluses of feed are brought back up to the mouth via the esophagus for cud chewing. The grinding of feed between the molar teeth further breaks down plant fibers. The rumen lining has finger-like papillae that absorb the volatile fatty acids resulting from the enzymatic breakdown of plant starch and fiber. These fatty acids— acetic, propionic, and butyric—provide a large percentage of the energy to meet the goat's nutritional demand. The microorganisms also play an important role in making Vitamin K and the B vitamins. The rumen begins functioning in a goat kid at around 3 weeks of age and usually reaches full capacity at 12 weeks. The capacity has a range of 12 to 28 liters, with 20 liters being maximum capacity in most goats. This varies depending on breed and ration fed[1].

The next compartment is the omasum. This compartment is the smallest of the four, but has a lot of surface area due to the multi-fold lining. This absorptive feature reveals its function: to absorb water and minerals prior to passing the rumen-digested feed to the abomasum or true stomach.

The abomasum has a similar function to that of a human stomach. The lining cells secrete hydrochloric acid and the enzyme pepsin. This enzyme breaks down proteins in the feed prior to moving the ingesta into the small intestine.

In young kids, prior to rumen development, milk is shunted directly to the abomasum through a reticular (esophageal) groove. The stomach of a young kid produces chymosin (rennin), which coagulates the milk proteins. This enzyme is extracted for cheese making, as well as now being synthesized commercially. The mature abomasum has a volume of about 4 liters.

The small intestine is well-packaged into the right abdominal side of the goat. The mean length of the small intestine of larger goat breeds is around 75 feet. Most of the small intestine is convoluted and fits neatly into its abdominal space. Absorption of most of the remaining nutrients occurs through the lining of the small intestine. A myriad of vessels, both blood and lymph, move amino acids, fatty acids, and sugars to be used by destined body cells. Digestive secretions from the pancreas, gall bladder, and the intestinal lining itself aid in the breakdown of remaining feed nutrients. The small intestine can be permanently damaged by severe coccidia infections, lessening the absorption capabilities of the lining.

The large intestine completes digestion of feed nutrients. The main function of the large intestine is to absorb water and some vitamins. The fecal pellets are formed in a coiled portion of the large intestine called the spiral colon, which has an interesting configuration of turns. A series of specific contractions form the classic pellets. The feces move through the final area of the large intestine into the rectum and are passed out through the anus.

Learning the gastrointestinal system and how it functions helps in understanding what goats eat and why. I encourage all goat keepers to take an opportunity to explore its complexity with an autopsy of a deceased animal. Beginners are advised to do this with a veterinarian or goat mentor to learn the basic parts and see the structural positions inside a goat, size comparisons of the compartments, and various system linings. Reading about or seeing pictures are one thing. Examining with your own eyes and feeling the lining textures with your hands enliven the lesson. This firsthand knowledge will help you understand your goats literally from the inside out.

1 *Goat Medicine, 2nd edition.* Mary Smith and David Sherman, 2009, p 379 Nutrient Requirements of Small Ruminants, NRC, 2007, The National Academies Press.

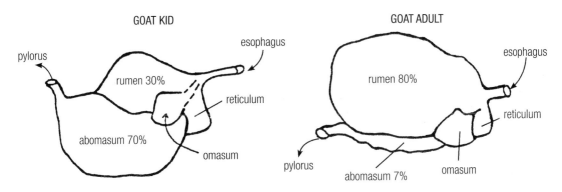

GOAT KID

pylorus

esophagus

rumen 30%

reticulum

abomasum 70%

omasum

GOAT ADULT

esophagus

rumen 80%

reticulum

pylorus

abomasum 7%

omasum

The gastrointestinal system of your goat. *Courtesy Barbara Carter*

Basic Feed Nutrients

The basic feed nutrients needed by goats are energy (carbohydrates), proteins, fats, minerals, vitamins, and water. Each nutrient is so important for the growth, maintenance, and production of a dairy goat; we'll take a brief look at each.

Energy (Carbohydrates) – Goats need energy first to support the rumen microorganisms, and then to support the goat itself. Carbohydrates are the main source of feed energy, though proteins and fats can also supply energy. Grain and plant starch and fiber are primary sources of energy. Lactating goats need a lot of energy in their feed. Often, body condition will suffer when the goat is in peak lactation. The animal simply cannot eat enough to meet the demand. Monitor these goats closely, watching for parasite burdens and adequate access to good forage. Change pasture/browse areas frequently so they have fresh forage, feed dry hay many times a day so they have access to as much as they need without wasting, and feed grains moderately. Energy demands at late-stage pregnancy are also high. Be sure to provide balanced feed during this period.

Cut hay bale twine at the knots to make the twine more easily reusable. A knot in the middle of a string can get in the way.

Protein – Protein is a building–block nutrient containing nitrogen. Goats need protein to grow their own bodies and those of their fetuses, as well as for health, maintenance, and milk production. Protein feeds are generally the most expensive and scarce in many locations in a browse area. There are high protein leguminous trees, such as black locust and leucaena, but limit sudden feeding of both of these plants due to mild toxins. Alfalfa, clover, vetch, and other low-growing leguminous plants also offer protein and can be pastured by the goats with careful planning; *ensiled* (preserved in a silo) with caution for proper anaerobic fermentation to avoid mold and listeria contamination; and cut, cured, and baled for dry hay. Pastures with pure alfalfa will be too rich for goats that are unaccustomed to it. Feed dry hay prior to turning animals into alfalfa pastures, and limit the amount of acreage and time they spend there. Goats do not readily eat fresh clover, so mixing a few calves with goats will make good use of a clover/grass/browse pasture, as the calves will enjoy the clover while goats eat grass and browse. Soy is often added to grain rations to boost the protein levels.

Fats – Though fats provide more than twice the energy of carbohydrates, there is little in a goat's diet, as plant material does not contain usable fats. Adding them to feed because of the high energy they provide is attractive, but ration fats must be limited and are best protected from degradation in the rumen. High fat levels in the rumen can cause difficulties with normal plant fiber digestion.

Supplement mineral needs with a goat-specific mineral mix.

Minerals – While energy and protein are important nutrients for growth, reproduction, milk production, and general health, minerals play a pivotal role in keeping all the systems working not just well, but working, period. Imbalances or missing minerals will throw off vital systems in a goat, sometimes with fatal results. There are fourteen essential minerals, both macro-minerals such as calcium, phosphorus, potassium, and magnesium; and micro-minerals such as cobalt, copper, iron, and zinc. Goats that are browsing fresh forage have increased intake of naturally available minerals. Their natural habit is to ingest a range of plant material, getting different nutrients with each bite. Fresh forages being cut and carried to goats should mimic this feeding style, bringing a variety for the goats to pick through. Minerals can be supplemented through goat-specific mineral mixes, general trace mineral salt blocks, and other supplements, such as kelp.

The supplements are important during winter months in cold regions where fresh forage isn't available. Retaining a 2:1 ratio of calcium to phosphorus is important. An imbalance can cause problems of calcium uptake with the start of lactation, as well as other issues.

Vitamins – Vitamins are also important for the overall function of a goat. Deficiencies often manifest as inefficiency and unrealized potential that go unnoticed and uncorrected rather than through any outward signs of the goat's health. Vitamins are either fat soluble or water soluble. Fat soluble vitamins include A, D, E, and K. Water soluble vitamins include the family of B vitamins and Vitamin C. The rumen microbes synthesize B vitamins. Vitamins are necessary in small amounts, though more research is needed to understand their roles and requirements.

Water – Water is considered a feed nutrient. As with other nutrients, without this vital intake, a dairy goat will not grow, produce milk, or maintain health. Water is a major constituent of nearly all living organisms. Milk is 87 percemt water; therefore milk production and general growth will be minimized without frequent access to water. Adequate water intake for male goats can reduce the incidence of urinary calculi formation. Sourcing clean water for your goats is vital. Many areas are facing water shortages, and in communities where hobby animals are not considered essential, water access may be limited. If you live in an area that is facing drought and water restrictions, verify water resources.

Safety first! Don't drink out of a hose on a farm. Hoses can be contaminated with manure, chemical residue, and pathogenic bacteria. Replace old, cracked, lead-free hoses and nozzles. Test water annually for coliform, nitrates, and nitrites.

Many good books and Internet resources give in-depth descriptions of the nutrients, how the gastrointestinal system breaks them down, and the body uses them. Anyone interested can spend days learning about each nutrient. This fascinating subject is still being researched and further understood as it applies to dairy goats.

Understanding the building blocks of nutrition allows you to knowledgably consider various feedstuffs for your animals. Use local sources, whether you are able to grow for yourself or buy from neighbors. A unique flavor of the milk and dairy products come from plants grown in your local soils and conditions. If you don't have affordable access to a variety of plants, utilize the resources you do have.

Eating Habits and Behavior

Goats are adaptable browsers. The constant in their feeding habit is variety. They thrive with the ability to be out in a browse area gathering goodness from local, seasonal plants. Goatherd them into a wooded area of maple, oak, poplar, and birch and watch the goats devour what they can reach. They will, though, want to move on, sometimes sooner than you do. They will move to a woods edge to munch bittersweet, poison ivy, brambles, Japanese knotweed, and grapevines.

Next they will move into a field and take bites of alfalfa, grasses, and weeds growing in the field. They will generally take a rest and begin digesting the plant material they have gathered and stored in their rumens. After some time of lying in the shade and chewing their cud, they will rise as a group and start foraging again.

This is great if you have a situation where they can simply free range without fear of poisonous plants (such as mountain laurel), roadways, neighbors, or predators. Otherwise you can use fences and plan some time for both woods browsing and pasturing in the fenced area. Set up your fence with a funneled opening to bring the goats from the woods into the desired pasture space. Make sure they have some shade in the pasture area and water if necessary. If no appropriate water source is available, you can train your goats to use a hog watering nipple. Retrofit a large water cube with a hog watering nipple, and place the cube

at a good drinking height for your goats. Train them to drink from the nipple, and then move the cube into your pasture area.

Note: Goats that walk a distance will expend energy. They will need to eat more to make up for that energy loss to sustain milk production levels. I find, though, that eating fresh forage, even with walking to browse, increases the quantity and improves the quality and flavor of the milk.

The urushiol oily compound of poison ivy can be on a goats' hair coat if they are browsing in a patch. Take note to wash well or avoid touching the animals if you are sensitive to poison ivy.

Determine browse plant types—shrubs, small trees—that will grow in your area and plant them for your goats to browse. Check to make sure you are not planting invasive species. Even with goats browsing, some plant species are aggressive growers and will crowd out native plants.

Goats will pasture on grasses such as orchard, bluegrass, timothy, Bermuda, and ryegrass. They do prefer a mix with leaf and twig stem fractions.

Goats happily browse in the woods.

Train your goats to access a hog watering nipple, which can be fitted to a water cube.

They will move from browsing to field edge.

Staghorn sumac (*Rhus typhina*) and brambles (*Rubus* spp.) are well liked. Herbs and wild plants such as nettles (*Urtica* spp.), asters (*Aster* spp.), rose (*Rosa* ssp.), and white Campion (*Lychnis alba*) are some favorite goat plants that grow in pasture areas.

Although this in many ways goes against natural goat behavior, set up small areas with electric net fences and permit the goats limited but concentrated pasture. Moving them to new feeding every 12 to 24 hours is ideal. Animals need to be moved when blade height is grazed to a minimum of 4 inches, as discussed above. Determine the size and frequency of moving pasture with consideration for precipitation, soil conditions, plant species, and plant growth rates. Rotating animals through a pasture system and providing time foraging in the woods and at field edges allows the goats to get a variety of plants and nutrients. Supplement with dry hay as needed in the barn or paddock area and grain at milking if necessary to achieve the animals' potential and your production goal. Grain can be a premixed concentrate, 14 to 18 percent protein, or a locally made grain mix. Mineral mixes and salt should be available at all times.

Goats can also be kept in smaller raised pens and brought forage, although this requires daily labor to gather feed and small equipment for handling it. This is a good alternative if you do not have an area for goats to browse. Be sure to select a variety of greens that goats like and watch their body condition, making adjustments to their feeding as needed.

Many goats are managed in a confined area, with an exercise paddock. Goats certainly can grow, make milk, reproduce, and maintain

Willows and Goats

I planted a willow (*Salix capra*) browse area in a damp lower field. This damp area is cool in the summer, and I like to put goats there in the heat of July and August. I don't like the goats eating grasses with their heads down in that area, as the damp field is more likely to host parasites larvae. Willows like a wet area, and goats like willows. Also, willow leaves contain condensed tannins, which may help lower parasite burdens. I planted willow branch cuttings in the spring and protected them with a wire cage that will remain on the seedlings to keep the goats from eating the bark and killing the trees. I currently limit the goats' time in this pasture so the saplings can develop more strength and leaf growth. Initial findings are that the goats like the willows and prefer to browse with their heads up. The goats will eat the grasses around the trees, but I am sure to move them out when the forage is down to a minimum of 4 inches, as parasite larvae normally travel 2 to 3 inches up a grass stem. Moving animals out of areas when grass blade height is eaten to 4 inches helps minimize parasite larvae intake. The willows give the goats additional forage in a cool area without letting them eat the grasses down too far.

Willow whip growing with protective guard.

Protect willow saplings and other new growth with strong wire protection.

themselves on a premixed standard ration, provided that they also receive plenty of forage, either as dry hay, silage, or green chop. Again, feed forages and grains that are locally available, understanding the needs for your goat's stage of life (i.e., growing, early lactation, late lactation, late pregnancy).

Goats, with their smaller digestive systems, prefer to eat several times a day. Mimicking their natural habit of foraging smaller amounts of feed and resting several times a day will help your goats thrive and increase production. This isn't always practical, especially if the family is working full time off the farm. Design a feeding

Above: Goats love nettles.
Left: Mixed goat grain is part of a balanced ration.

Body Condition

Body condition scoring is a monitoring tool for assessing the body's current state of energy reserves. The Body Condition Score (BCS), which ranges from 1 (extremely thin) to 9 (extremely obese), can be used to make changes to management and feeding practices. A dairy goat's body is in flux during lactation and pregnancy. A good energy reserve is needed to carry a goat through her pregnancy and into the first weeks of lactation. Too much reserve (fat) causes problems with pregnancy toxemia and other metabolic issues. Insufficient energy reserves leave a doe struggling to meet her and the fetus' energy demands. This often results in weakness at kidding and a lower milk production throughout that lactation. The proper body condition at particular stages throughout a lactation year is important for health, increased milk production, breeding success, and ease of pregnancy and birthing.

BCS is a hands-on assessment. You are feeling specifically for the bone structure, attached muscle, and fat. Start by feeling the goat's spine at the base of her neck. Follow down into the shoulders and further down to the sternum. Go back to the spine and run your hands along the vertebrae. Feel down along the ribs, back up to the spine, through the hook, and back over the rump to the pin bones. Feel over the tail head. This assessment should give you an overall understanding of the amount of body fat and condition of the muscle. Assign a number based on the following chart and enter this in the individual animal record, along with the date. Ideally an assessment should be made three to four times a year, and in conjunction with FAMACHA. Score animals before breeding, in the last 3–4 weeks of pregnancy, and 6–8 weeks into lactation. Ideally a moderate BCS of 4 to 6 should be maintained through the lactation year, with a higher score of 5 to 6 at the end of pregnancy and a lower score of 4 to 5 at 6 weeks into lactation. Make adjustments in nutrition and management to achieve this healthy body condition.

A low BCS (1 to 3) can be indicative of a high parasite load, poor nutrition or absorption of nutrients, and health concerns. Animals with a high BCS (7 to 9) should be managed with more activity and a change in their feeding routine. Remember that all feeding changes need to happen gradually. A dairy goat that tends to put excess fat on her body is generally not a highly productive goat. BCS trends can be used in conjunction with your goals when making decisions at breeding and culling time.

schedule that works for your lifestyle, striving for three to six feedings per day.

Available feeds for goats are generally locally grown. Hay made with local grasses and legumes is available in a wide range of quality. Ask your provider for an analysis of the particular hay you are buying, or submit a sample yourself. Grain companies, extension agents, and dairy herd improvement associations (DHIA) will often submit hay samples to a licensed laboratory for a small fee. Ask for help understanding the results and how this applies to feeding your goats. Processed feeds, such as pasture cubes and hay stretcher, are available in many locations. These products help in times when hay is not available.

Body Condition Score

LOW	MODERATE	HIGH
1 Emaciated; near death, very thin, weak with atrophied muscle	4 Slightly thin; some ribs visible; thin flesh covering hooks and pins	7 Fleshy; frame not visible, spine felt with firm pressure, hooks/pins smooth
2 Extremely thin; not as weak as BCS 1; ribs and spine visible	5 Balanced; ribs and spine felt smooth with flesh, muscles obvious	8 Fat; ribs cannot be felt, obese, tail head cavity becoming fatty
3 Very thin; ribs visible, spine prominent, no fat cover under skin	6 Slightly fleshy; ribs smooth, not very visible, spine felt with pressure	9 Very obese; entire body covered with extreme amounts of excess fat

Genetically modified plants are dominating the animal feed world. Opposition to these feeds creates opportunities for exploring other options. Organic feeds are available on a limited basis, depending on your locale. Buying in bulk with other animal keepers can reduce the costs.

The bulk of nutrients should be obtained with forages, fresh, fermented, or dried. Offer ensiled (fermented) feed with caution for mold and listeria contamination. Grains, either commercially processed or locally mixed, can be added to balance the nutritional needs of your goats. Below is a basic guideline for nutritional needs based on various production stages. Specific information on formulating rations is beyond the scope of this chapter. Langston University has developed the Langston Interactive Nutrient Calculator (www.luresext.edu/goats/research/nutr_calc.htm), which is useful for figuring goat rations. Another useful tool is a computer software CD developed at University of California at Davis (animalscience.ucdavis.edu/extension/software/capricorn/).

Watch your animals' body conditions and adjust feed accordingly. Work with your goat mentor, veterinarian, and local agricultural extension to learn more about feeding. Feeding goats is a science and an art. The science is in learning how the body works and the chemical breakdown and use of nutrients. The art is in observing your animals' body conditions and health, and adding feeds that will improve their lives. Too much of a good thing can be too much, so be sure to teach children and visitors to go slow and easy with treats. Raisins, peanuts, popcorn, black oil sunflower seeds, apples, pears, berries, carrots, bananas, and garden greens are nutritious treats goats enjoy.

There is currently a surge in finding alternative feeds. Dairy barley fodder, which sprouts barley seeds growing 6 to 7 inches, makes a good feed alternative. Interestingly, this feed was mentioned in Dr. Frank Morrison's *Feed and Feeding* from the 1950s. Barley fodder did not gain much popularity back then, as grain was inexpensive and readily available. Other feeds being explored are moringa, spent grains from brewers, and sainfoin.

Basic Guideline for Nutritional Needs

	Daily kg	Dry Matter[5] % BW	Energy [6] ME Mcal/d	Protein[7] MP g/d	Ca[8] g/d	P[9] g/d
Mature Doe[1], Maintenance	1.35	2.25	2.58	61	2.4	2.0
Mature Doe, Breeding	1.49	2.48	2.84	67	2.6	2.1
Mature Doe, Early Gestation, twins	1.64	2.73	3.14	100	6.3	3.7
Mature Doe, Late Gestation, twins	1.69	2.82	4.05	143	6.4	3.7
Mature Doe, Early Lactation, milk = 11#/day	2.58	4.30	7.40	364	16.5	9.8
Mature Doe, Mid Lactation, milk = 9#/day	3.00	5.00	7.17	318	17.1	10.3
Mature Doe, Late Lactation, milk = 6#/day	3.08	5.13	5.89	252	17.2	10.4
Mature Buck[2], Maintenance	2.28	2.28	4.36	97	3.7	3.2
Mature Buck, Pre-breeding	2.51	2.51	4.79	106	4.0	3.5
Growing Kid[3], Doeling	1.02	3.41	2.47	83	5.6	2.9
Growing Kid[4], Buckling	0.96	3.21	3.01	97	6.7	3.3

[1] Mature doe 132#, 60 kg

[2] Mature buck 220#, 100 kg

[3] Growing kid, doeling, 66#, 30 kg

[4] Growing kid, buckling, 66#, 30 kg

[5] Daily dry matter, kg and percent body weight

[6] Metabolizable energy

[7] Metabolizable protein

[8] Calcium

[9] Phosphorus

Nutrient Requirements of Small Ruminants, National Research Council. 2007.

Goats love spent brewer's grain as much as brewers love the beer made from it!

Research these interesting feeds to see if they fit into your goats' nutritional program and are available in your area. As with all new feeds, introduce alternative feeds slowly, allowing the rumen microorganisms to adjust.

Wear light-colored or white clothing when working with hay or any chore on a hot summer day.

Feed Quantities

Goats will eat 3 to 5 percent of their body weight in dry matter (DM) roughage per day. Thus a 120-pound goat will eat between 3.6 and 6.0 pounds of DM roughage. Basically speaking, dry matter is the weight of a feed with water removed. Dried hay is around 90 percent dry matter. Fresh forage is about 10 percent dry matter, so a goat needs to eat 36 pounds of fresh forage to get 3.6 pound of dry matter.

Fresh forage: 3.6 lbs. (DM required) /
.10 (10% dry matter)
= 36 pounds (fresh forage) as fed

Dried hay: 4.5 lbs. (DM required) /
.90 (90% dry matter)
= 5 lbs. (dried hay) as fed

If you choose to feed a 16- to 18-percent protein grain mix for a milking doe, the basic recommendations are one pound for maintenance and one pound per every quart of milk being produced per day.

Colorado Intermountain Goat Grain Mix

40% rolled flaked corn   20% rolled wheat
20% rolled barley   10% soybean oil meal

The remaining 10 percent includes liquid molasses; a vitamin A, D, E supplement; and other nutrients. Beet pulp can be added. The grain ration should not include more than 20 percent of any one cereal grain and total cereal grains should not exceed 40 percent.

Estimating Weight with Heart Girth Measurement

Inches	Pounds	Inches	Pounds
21 ¼	35	31 ¾	97
21 ¾	37	32 ¼	101
22 ¼	39	32 ¾	105
22 ¾	42	33 ¼	110
23 ¼	45	33 ¾	115
23 ¾	48	34 ¼	120
24 ¼	51	34 ¾	125
24 ¾	54	35 ¼	130
25 ¼	57	35 ¾	135
25 ¾	60	36 ¼	140
26 ¼	63	36 ¾	145
26 ¾	66	37 ¼	150
27 ¼	69	37 ¾	155
27 ¾	72	38 ¼	160
28 ¼	75	38 ¾	165
28 ¾	78	39 ¼	170
29 ¼	81	39 ¾	175
29 ¾	84	40 ¼	180
30 ¼	87	40 ¾	185
30 ¾	90	41 ¼	190
31 ¼	93	41 ¾	195

Feeding Equipment

Goats are known for picking through forages and playing with feed buckets. This behavior often results in wasted feed and unnecessary expense. Place hay at a height that requires the goats to reach for the forage. Another idea is to use a key hole or slanted slats in the hay feeder to minimize the amount of hay that is pulled out. The goats will eat the hay at the feeder and remove their heads without pulling at the hay when they have finished eating. Vertical slatted hay feeders are better than horizontal slats.

Goats like to stand on a step to eat. The problem with having a grain tray under a hay feeder is the goats often stand on the edge of the grain tray to reach the hay. Manure and dirt are left in the grain tray when the animals stand on the edge.

Young goats jumping into hay racks may seem cute at first, but the hay will quickly be spoiled by kids urinating and defecating on the feed. Discourage this habit with a simple wooden frame that fits into the hay feeders.

Be sure to keep feed racks, grain trays, mineral feeders, and water buckets clean of manure and debris.

Grain buckets used at milking time can become a play toy for a bored goat. Find clever ways to deter this habit, such as clipping the bucket to the stand with a carabineer or simple fence clip.

Mineral feeders and tubs should also be kept at a proper height so manure and urine do not soil the supplement. Keep these dry materials out of the elements.

Waterers and water buckets need to be cleaned daily or whenever they become soiled. Water buckets containing feces, hay, insects, fowl droppings, and other contaminants need to be dumped and refilled. Scrub buckets with hot soapy water and a brush as needed, or at least several times a week. Bowls of automatic waterers need to be kept cleaned and in good working order. Perform annual maintenance checks of heating elements, wiring, and water flow parts for automatic waterers, preferably in the fall in colder areas. No one wants to mess with fixing a waterer in the deep freeze of winter. Frost-free hydrants properly installed make easier work of watering animals in the winter, if automatic waterers are not practical.

While visiting other goat farms, check out their feeding equipment to determine if something they do will work for you. Be clever and save money by reusing items such as pallets and used lumber. Make sure anything

Estimating the body weight of goats

To determine the weight of a goat, measure the goat around the heart girth. Pull the tape lightly.

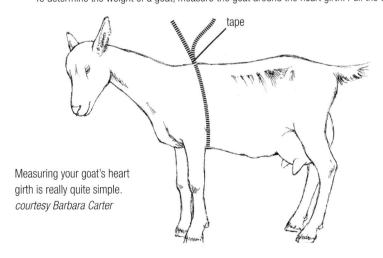

tape

Measuring your goat's heart girth is really quite simple.
courtesy Barbara Carter

you use does not have exposed nails, screws, sharp metal edges, or other hardware, as well as wood paint or other wood treatments that can harm the goats.

Feeding goats can be simple, quite involved, or anywhere in between. Feeding plans should follow your goals. Having a few backyard goats making delicious and nutritious milk can require simple feeding of fresh forage, local hay, and a little grain if needed. Breeding genetically superior animals or milking commercial production goats requires more knowledge and fine-tuning nutritional demands and feed quality and quantity.

Assess the body condition of your animals, the production quantity and quality, overall health, and breeding/pregnancy condition. Adjust the feeding protocol to keep your animals in good condition and thriving with strong immune systems. A sound plane of nutrition helps to build a robust, long-lived, and productive goat. Provide good feed, practicing both the science and art of knowledge and observation. If possible, allow your goats to forage and browse in fields, woods, and transitional areas. Bring them cut forage as an alternative if browsing areas are minimal. Carefully adjust their feeds—remember you are feeding the rumen bacteria. Be cautious of poisonous plants and harmful situations, such as those involving predators or bad weather. Minimize stress, work in a routine, and limit exposure to parasites and infectious organisms. Spend time just having fun your goats, and capture the joy of watching them eat, knowing that a balanced variety of nutritious feed is vital to their growth, health, and productivity.

Fresh goat forage being gathered in Grenada. *The Goat Dairy*

Left: Goats eating grain in a milking parlor. *Danielle Mulcahy*
Below: Does enjoy the whey from cheese making.

Chapter 4

Management

Manage (*verb*): To have control of; to take care of or make decisions about. Managing dairy goats, while juggling other aspects of life, is challenging. Plan, prioritize, organize, maintain equipment, keep supplies stocked, and train help. Wiggle when necessary, laugh when you can, and cry when all else fails.

As a longtime goat farmer/philosopher once told me, "Everyone has their wild cards. How you play your cards is what makes the difference." Identify *your* wild cards and learn how to play them to your advantage when setting up or improving your dairy goat herd. At the same time, identify your weaknesses and find either a mentor to teach you or skilled people to do these tasks. The joy of a challenge is learning new skills.

This management chapter covers a range of topics; some will undoubtedly be new skills for you. The topics are connected by the central theme of caring for your dairy goats.

Maternal bonding with newborn kids is important for the herd's well-being.

Goat bells galore at a dairy farm in the Canary Islands.

Understanding Goat Behavior

"GOAT-titude" opens the door to caring for and understanding goats. Their behavior is a bit unique in the farm animal world. They are herd animals, yet quite independent in their actions and personalities. Bonds between a doe and her kids or goats born in the same year and raised together are generally strong.

I see eight-year-old goats that still gather and lie down together. One goat encourages another during kidding, or stops a fight between others. Their communal mentality of survival is interesting to observe. Most goat owners, regardless of their herd size, say their goats' relationships and how they interact are noticeable. Individually, goats are strong minded. They are downright bullies at times. The alpha dominance is alive and well in the goat world. Social hierarchy is frequently challenged and changed. Curious, though, that the alpha goat is often not an animal that leads the herd. This role often goes to the most inquisitive. Regardless of who leads, the rest will generally follow, often at their own pace

and discretion. Put a bell on your lead goats and listen to the soft melody as they go to and fro with browsing.

Goats like routines. Teach a milking stand routine and most goats will oblige eagerly. As discussed in the previous chapter, set up your pens to make the movement flow easily for the animals. Use gates to make small holding pens. Minimize stress, especially at milking time. Young goats quickly learn to follow and will make a game out of going for walks. Teach them to follow to a call. With repetition and acknowledgment, your goats will come along and go where you want.

One management tool that works well for moving goats is a stick—nothing fancy, simply a stick. Teach the goats to follow you when you are holding the stick in the air, and to stop when you put the stick out horizontally. When you simply walk with the stick by your side, the goats are free to roam. This training takes some persistence and patience, but when the goats learn, they will teach the newcomers to follow your lead.

To move a stubborn goat, turn the animal so its tail faces the direction you want it to move. You can drive the goat backwards as they can't "put on brakes" by bracing their front feet.

As a group, goats have a sense of loss with the death of a herd mate. At the same time, they have no patience for a sick or weak animal. The herd survival mentality is instinctive. Be careful incorporating a healing animal back into the herd. It will have to find its place in the hierarchy again.

Dairy goats are strong, adaptable animals. Encourage that tenacity, that character-building strength in your goats. Provide them with their basic needs and they will provide you with plenty. Remember they are food-producing farm animals that like to play, be challenged, and provide companionship.

Keeping Records

Records are an important management tool. To make decisions, you need information. Today's technology allows you to easily create databases. Keep a barn log. Decide what works for you. I enjoy the simplicity of a notebook and pen kept in a three-ring binder for my barn log, recording animal health information, feed changes, animal sales, and other pertinent details. Information is separated into topics such as daily notes, breeding, birthing, kid care, mastitis, nutrition, medication/vaccinations, sales, and miscellaneous. This is where I write information while I am doing the job, healing the goat, or simply making notes for future reference. I will take information and transfer it to individual animal records as needed.

Raising and selling breeding stock requires good recordkeeping. Buyers want to know information on milk records, birthing statistics, general health, vaccination and deworming schedules, and family history. Keep these records updated to avoid a rush to find records when you have a buyer at the door. See Appendix 3: Dairy Animal Record Sheets at the back of the book to create your own database using spreadsheets, and/or incorporate

electronic milk statistics with individual and herd data from a DHI service.

Financial records are important at any level of goat keeping. Making a profit from your dairy goats requires a strict schedule of creating and tracking budgets, figuring income and expenses, and following individual goat's milk production. Develop a weekly, monthly, quarterly, and annual system for doing these tasks. If financial recordkeeping isn't in your deck of wild cards, outsource this job. Listen to your advisor, add in your own knowledge, ask questions, and make informed decisions. Learn from mistakes and keep moving in the direction of your goals. Reevaluate and make changes. The most important point is to not ignore the facts or turn them into something else so you can feel better. If your goats are not generating the income you want, find out why. Can you work harder? Go outside your comfort zone with marketing or sales? Were your goals realistic with your surroundings, skill level, and resources? Be truthful and kind to yourself and your family and friends.

Identifying Goats

Animal identification is an important aspect of dairy goat management. Individual animals are identified for recordkeeping, health issues, parasite evaluation, milk quality and quantity analysis, genetics, and fun. The type of identification often follows your goals. Registered animals must be tattooed for permanent identification. Larger herds often have a tattoo or tagging system. Smaller backyard herds will often simply have barn names.

Write identification on a Dairy Animal Record Sheet. These sheets also include other pertinent information, such as birth date, parents, vaccination, deworming and health dates and issues, and breeding and birthing dates. Photographs can be taken and attached to this log.

Barn names allow for personal attachment to your dairy goats. This is important as the animals are handled every day for milking chores and enjoyable associations are established. Barn names are created by many means. One example is a breeding line that uses names with the same first letter or theme (e.g., Rose's offspring will be named after flowers).

Collars are useful for handling and identification.

Another idea is to assign names by the first letter matching the required ADGA tattoo letter of the year (all kids born in 2015 names begin with letter F).

Neck collars with identifying tags are useful, especially in larger herds or at farms with multiple people tending the animals. Collars can be plastic breakaway chains, cloth buckled collars, or metal chains. Management should determine the type of collar used. The plastic breakaway collars are beneficial for animals that are out in browse or pasture as they may otherwise get caught in branches and choke. Collars and tags can get stuck in headlock or stanchion milking systems. Cloth buckled collars are not easily lost and make handling the goat simple. Tags can be purchased in many colors and styles, blank, numbered, or custom made. Select a system that works for you.

TATTOOING

This is the most permanent means of identification. Dairy goats are tattooed in the ears, except for the LaMancha breed, which is tattooed in the tail web. ADGA requires tattoos for registered animals. Registered herds are assigned an ADGA tattoo herd sequence. This tattoo is to be placed in the right ear/tail web for all animals initially registered to that herd. The left ear/tail web is tattooed with the designated year letter (2015 - F) and serial number in order of birth (F1, F2). Another method, if not registering with ADGA, is to use the last digit of the year and order of birth (i.e., 501, 502 would be the first and second kids born in 2015). Reading tattoos is aided by cleaning the ear and shining a flashlight from behind it. See the video at vimeo.com/user29715351.

Tattoo pliers, symbols, and ink—essential identification tools for farmers.

Tattoo pliers with symbols secured in place.

BASIC TATTOOING STEPS

1. Prepare a tattoo station with tattoo pliers, symbols, ink, alcohol, gauze or swabs, worksheet, and restraint.
2. Arrange the first set of symbols to be used in the pliers and pierce a piece of paper to check for accuracy.
3. Retrieve and restrain the animal. Clean its ears/tail web with a swab soaked in alcohol.
4. Smear ink on the skin. Place pliers parallel to and between veins and/or cartilage of the ear/tail web.

5. With a strong, swift motion, pierce the area. Most pliers come with a release pad that helps separate the ear/tail web from the needles. Apply additional ink and rub it into the imprint with cotton gauze or a swab for 10 to 15 seconds.
6. Record the tattoo and animal information on a worksheet for future transfer to individual animal records.
7. Clean symbols and pliers with alcohol and water. Dry thoroughly before storing.

EAR TAGS

Ear tags are a useful means of identification that are easy to read and generally more permanent than neck collars. The downside of ear tags, particularly in goats feeding on browse, is the risk of the tag getting snagged on brush and ripping the tag from ear, causing a tear. Small metal tags, such as those used for lambs, can be useful for kids, though these are more temporary and not designed for mature animals.

As with collar tags, plastic ear tags come in many sizes, colors, and styles, blank, numbered, and custom made. Select a size and variety that works for your herd. Ear tags generally remain in the ear for life, so consider the growth and future needs of the dairy goat.

BASIC EAR-TAGGING STEPS (USING PLASTIC TWO-PIECE EAR TAGS)

1. Prepare ear tag station with the applicator, tags, worksheet, and restraint.
2. Place tag pieces in the applicator, making sure the pin piece aligns with the hole of second piece for proper tagging. Some applicators are designed so the "flag" of the tag lies inside the applicator and does not touch the ear until the tagging is complete.
3. Retrieve and restrain the animal. Clean ears/tail web with a swab soaked in alcohol.
4. Place the applicator with tag in the center of the ear, between the skull and the end of the ear, avoiding the ear veins and cartilage. Quickly squeeze the applicator and snap the tag in place.
5. Record information on a worksheet for future transfer to the individual animal record.

Don't worry—tattooing the tail web of a LaMancha kid is not as painful as it looks.

SCRAPIE TAG

The USDA, in an effort to reduce and eliminate the disease scrapie, has a flock/herd certification program that requires all sheep and goats be tagged with an assigned flock/herd identification number. Scrapie is classified as a transmissible spongiform encephalopathy (TSE) that is fatal and untreatable. It is a degenerative disease affecting the central nervous system. Luckily, the incidence of scrapie in goats is very low. Contact your local USDA/APHIS office or www.eradicatescrapie.org for more information on this program. Some states follow federal regulations for scrapie identification, while others have additional state regulations. Free USDA ear tags can be requested from the local APHIS or state veterinarian office, or specific scrapie tags can be ordered from an approved tag company. Moving goats across state lines requires identification and veterinary health certificates as called for by state and federal agencies. Contact your local agricultural office for more information about moving goats across state lines. Follow the basic steps for ear tagging to insert the scrapie ear tags.

MICROCHIPPING

Microchip implants are an effective means of identification. This simple device is basically a small computer "biochip" implant, inserted under the skin with a hypodermic syringe. The implant system consists of two parts: a transponder (the actual implant) and a reader. The reader sends a low-frequency radio signal to the transponder. The transponder biochip stores a unique identification number, generally ten to fifteen digits long. The reader, communicating with the biochip at a distance of 2 to 12 inches, receives back this unique identification number.

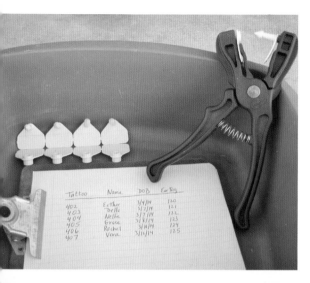

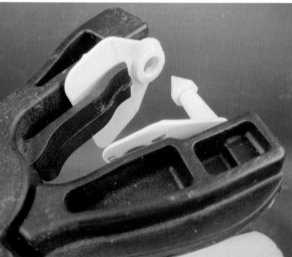

Top: Keeping a notebook handy makes for easy identification and organization of newly tagged goats.

Above: An up-close shot of the applicator.

Disbudding to Prevent Horn Growth

Disbudding, or destroying horn bud cells by applying heat so the horns will not grow, is a recommended technique to eliminate horns from your goats. This is done in goat kids a few days to a few weeks old and is generally a simple procedure. Dehorning is removing the horns on older animals and is more invasive and complicated. Disbudding is recommended over dehorning.

Your goals and management decisions will help determine if you should disbud your goat kids. Benefits of goats without horns include

ease and safety of goat handling for milking and health concerns, no injuries from goats fighting with horns, and decreased risk of animals being stuck in branches and fences. Most goat shows require animals to be hornless. Horns are beneficial to help goats cool in hot climates and defend themselves against predators. Stately horns, especially on bucks, are really quite attractive. Most goat keepers around the world keep horns on their goats, with the United States being an exception.

Polled goats, which are naturally hornless, carry a gene that causes reproductive tract problems. See Chapter 6, Breeding, for more information on this condition, and be aware of the genetic potential if you want to work with polled goats.

Male goat kids should be disbudded by 3 days of age, females by 2 to 3 weeks. Allow the horn buds to pop through the skin before burning. Disbud older kids with discretion and experience. See the video at vimeo.com/user29715351.

BASIC DISBUDDING STEPS

1. Prepare the disbudding station with a burning iron, extension cord, scissors, wire brush, thick cloth gloves, restraint box, and a pain-relief product (optional).
2. Plug in the burning iron, place it on a non-flammable surface, and allow time to heat, generally 5 to 10 minutes, depending on your equipment. Check the iron by dropping water on the hot surface. The bead of water should sizzle off instantly.
3. Retrieve the kid and place it in the restraining box. You may need a booster towel if the animal is smaller than the support board. Close the lid and sit on the box, straddling the contained kid.
4. Snip hair from around the horn buds. Hold the goat's lower jaw with one hand to help stabilize the head and apply slight upward pressure while the other hand applies the hot iron. Apply the hot iron directly and purposefully around one horn bud. Hold, applying pressure for 8 to 10 seconds, counting aloud. Remove the iron and examine the bud area. There should be a completely burned circle, copper in color.

5. Reapply hot iron for 5 seconds if the circle is not complete. Angle the iron slightly to burn any area that needs attention. Remove and look again. Repeat with caution to complete the burn.

6. The kid may scream, jump, and toss his head while burning. Hold firmly and complete the task as quickly as possible.

7. Allow the iron to reheat; this should only take a minute. Calm the kid. Repeat with the other horn bud, making sure you have a complete ring around the bud area. The few seconds of hurt is over very quickly, and in minutes, the kids are back to jumping and running.

8. If you choose, spray the treated area with a pain-relief product.

9. Scratch the wire brush over the disbudding iron end to remove any residue.

10. Allow iron to reheat between goat kids.

11. Record information on a worksheet for future transfer to the individual animal record.

Castrating Male Kids

Castrate male kids according to your goals. This procedure is generally performed on males with undesirable genetic potential, and to avoid strong odors and lessen the dominant attitude for the purpose of having a pet.

If raising male kids for meat, consider the customer's preference for meat from neutered or intact males. Castration at 3 months of age can affect the growth performance of kids fattened for 5 to 6 months in regards to mean carcass lean and fat percentages. Meat from castrated kids recorded a milder smell compared to meat from intact males (Zamiri et al. 2012).

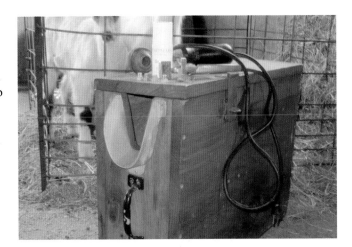

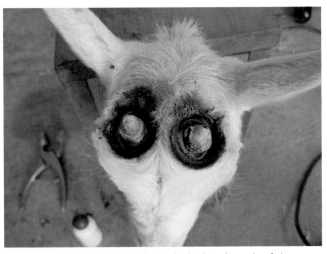

Top: A restraining box makes tagging easier (and much, much safer).
Middle: Burning the horn buds will help ensure the long-term health of your goats.
Above: A successful burn will soon be forgotten by your capricious companion, and he'll be healthier for it.

Castration is the neutering of males by removing the testicles or separating the connection of spermatic cords from the testicles.

Males can pass fertile sperm by 3 months of age, so any intact males you keep need to be separated from females unless they are intended for breeding. Plenty of young females have suffered or died due to birthing when too small because the goat keeper said, "I'll separate next week." Sell or castrate males unwanted for breeding stock or specific meat sales, or separate them in pens with good fences before a disaster occurs 5 months later.

There are three main methods of castration. Regardless of the method chosen, a tetanus anti-toxin shot is recommended right before performing the procedure. Animals can also be given a Clostridium CD/T vaccine according to a vaccination schedule of 4 weeks and a booster at 8 weeks. Follow meat withdrawal times for animals being sold. Recommended age of castration is variable, from 3 days to 4 months, as long as the testicles are descended into the scrotum. Generally, younger animals experience less trauma and recover faster. A potential disadvantage of early castration is that resulting wethers may experience a higher incidence of urinary calculi (bladder stones that can interfere with urination). Local anesthesia may be required for older animals. Castration is best performed prior to insect season, if possible. Spray for flies as needed. Beginner goat owners are *strongly encouraged* to seek assistance from veterinary professionals or experienced farmers when castrating. See the video at vimeo.com/user29715351.

CASTRATION METHODS

Knife Method – This open-wound method is often performed on kids that are 2 to 4 weeks old. There is a risk of infection and tetanus.

1. Prepare a castration station with a sharp knife or razor blade that has been sterilized, alcohol, iodine (7%), soap and water, a clean cloth, and a restraint.

2. Retrieve and restrain the animal. The kid is best placed on his rump with legs held and scrotal sac exposed.
3. Wash hands and the scrotal sac with soapy water, dry thoroughly, and dip the scrotal sac in alcohol or iodine.
4. Make an incision 2 to 3 inches long on each side of the scrotal sac and push testicles out through the opening. The cuts should be low to encourage drainage of any blood or fluid that may accumulate in the scrotum after the procedure. The testicle's white, shiny tunic may also need to be cut.
5. *Gently* pull the testicles until the spermatic cord breaks (do not cut). Alternatively, the cord can be scraped with a scalpel blade or tied.
6. Dip the scrotum in iodine.
7. Release the kid and watch it closely for several days.

Emasculatome or Burdizzo Method
This bloodless method is generally used with kids that are 6 to 8 weeks old. There is less risk of infection and tetanus. Check the animal for atrophying testicles 4 weeks later to make sure the procedure was successful.

1. Prepare a castration station with an emasculatome (such as Burdizzo), soap and water, a clean cloth, and a restraint.
2. Retrieve and restrain the animal. The kid is best placed on his rump with legs held and the scrotal sac exposed.
3. Wash your hands and the scrotal sac with soapy water. Dry well with a clean cloth.
4. Push the testicles into the bottom of the scrotum and feel for the spermatic cord and blood vessels between your fingers. Place the emasculatome jaws onto the upper scrotum, with the cord and vessels of one side of scrotum in the jaws. Squeeze the emasculatome with slight pressure at first and then steady, firm pressure. Hold for 10 to 15 seconds.
5. Open jaws and move the emasculatome down ½ to 1 inch and repeat, making sure cord and vessels are in the jaws.

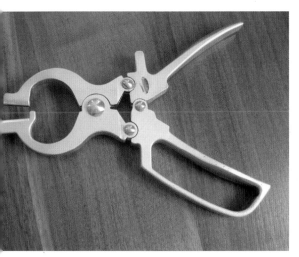

An emasculatome is an important tool for keeping your herd in check.

Gentle but steady pressure makes for a successful elastrating procedure.

6. Repeat on the opposite side of scrotum to separately crush cord and vessels on both sides. Be careful not to crush the middle of the scrotum, and avoid damage to the urethra.
7. Make sure the spermatic cord is between the jaws before and after crushing. The testicles will shrink and harden after 3 to 4 weeks.

Elastrator Method – This bloodless method is generally applied to kids fewer than 3 weeks old. This is the least expensive of all castration methods, but it carries the risk of tetanus and possibly increases urinary calculi difficulties. This method is no longer recommended by many in the profession of veterinary medicine or dairy goat husbandry, mainly due to the inhumane process of constriction to atrophy the testicles.

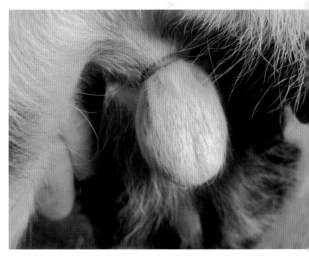

Castration bands work, but increase the chance of tetanus and other urinary tract difficulties. Many veterinarians and dairy goat farmers no longer recommend this form of castration.

1. Prepare a castration station with an elastrator, elastrator rings (heavy rubber rings), soap and water, a clean cloth, and a restraint.
2. Retrieve and restrain the animal. The kid is best placed on his rump with legs held and scrotal sac exposed.
3. Wash your hands and the scrotal sac with soapy water. Dry well with a clean cloth.
4. Place a rubber ring on the prongs of elastrator, turning the tool so that the prongs face the kid's body. Squeeze the elastrator to expand the ring and gently slide the rubber ring over the scrotum, manipulating *both* testicles into the bottom of the scrotum. Release the ring, being careful to avoid banding the rudimentary teats and keeping both testicles in the body of scrotum.
5. Carefully remove the ring from the prongs and slide the elastrator away from the scrotum.
6. The testicles will shrink and harden after 3 to 4 weeks.

Trimming Hooves

Hoof trimming is essential in most dairy goat herds. Dairy goats are generally fed higher protein rations, which promote hoof growth, and forage in areas with little gravel to wear hooves. Routine hoof trimming makes the task much easier. See the video at vimeo.com/user29715351.

BASIC HOOF-TRIMMING STEPS

1. Prepare a hoof trimming station with hoof trimming shears; a hoof plane; a sharpening tool; powdered garlic, blood stop, or antiseptic powder; and a restraint.
2. Retrieve and restrain the animal.
3. Lift a front foot and bend the goat's leg, resting her knee on your thigh.
4. Examine the hoof and determine how best to trim it, sighting a parallel to the hair/hoof line.
5. Remove dirt from the hoof, being careful to not poke the soft tissue with the point of the shears.
6. Trim away overgrown hoof.
7. Carefully cut the heel to same level, creating a flat surface that is parallel to the hair line.
8. Trim any growth between the toes.
9. Flatten and smooth the surface.
10. Apply powdered garlic (not garlic salt!), blood stop, or antiseptic powder to any areas that are overcut and bleeding.
11. Continue working around the goat, resting its thigh on your bent leg while lifting the rear legs.

BEDDING AND CLEANING THE PEN

A clean and well-bedded pen is important for dairy goats. Bedding materials can be sawdust, shavings, straw, non-moldy construction hay, or other materials. Find what is available locally and inexpensively. Bedding should provide comfort from a hard surface such as concrete or cold in northern climates. This layer of protection also keeps the udder clean from accumulating urine and manure, and minimizes bacterial infections that can cause

Regular hoof trimming will also help ensure the longevity and happiness of your herd.

mastitis and other health problems. The importance of cleaning the pen regularly is multi-fold. Primarily, a clean pen encourages better overall health of dairy goats, reducing mastitis, respiratory issues, parasite burdens, hoof rot, and other illnesses. Pest populations, such as those of flies and rodents, are greatly lowered in pens that are cleaned often. Goats are happier when their pen is freshly bedded— who doesn't like the feel and smell of clean bed sheets? General cleanliness of the pen and surrounding animal area is encouraging for the sake of visitors and neighbors. Remove trash and unnecessary or broken items.

During the cold months, a bedding pack provides insulation from the cold while generating heat from within. Put down a 6-inch-thick layer of sawdust. When this becomes soiled, top it with a thin layer of sawdust, straw, hay, and/or shavings. Repeat this as necessary. The pack will pull urine and moisture down, while insulating against the cold. Be prepared to remove the pack and put

down clean bedding before kidding season. This will be an involved job, particularly if you don't have a tractor or concrete floors. Consider renting a skid steer for the day and hiring some local teenagers.

 Pocket knives are a best friend at least once every day.

Controlling Pests

As mentioned above, regular cleaning and bedding will help keep fly and rodent populations down. Fly tapes, traps, and beneficial predatory wasps work to reduce flies. Keep grain covered, sweep aisles of debris, and hose and wash the milking area to eliminate milk residue. You should not smell milk when you walk into your barn. Place lids on trash receptacles and remove full bags regularly. Trap rodents, and keep a barn cat or two. Be careful using rodent bait, as this can poison an inquisitive goat.

A maremma-cross guardian dog in the field watches the dairy goats.

Neighboring dogs and natural predators can cause problems with a herd of goats. Good, strong fencing can help keep animals out of your pen and pasture areas. Consider getting a livestock guardian animal, such as a dog, llama, or donkey, if there is a threat to your herd. These animals provide real security as well as a relief from worry. Consider the expenses associated with these animals, including feed and veterinary visits. They easily earn their keep, with 24-hour protection for the herd. Our livestock guardian dog also lets me know when visitors are around.

Speaking of visitors, while they are in no way pests, there are some times when visitors need to be controlled. We welcome visitors to our farm both for tours and cheese sales. People stop in unannounced, wanting to visit both the animals and the farmers. I enjoy a break and stop to talk for a few minutes, getting back to work as needed. Set a schedule for days to permit visitors. Give tours by appointment only. Be polite and sincere. Allow visitors to enter your barn only with your permission.

Electric fences and other farm hazards need to be clearly marked. Discourage people from bringing their dogs to your farm and do not allow them to let their dogs run unleashed, especially if you have guard animals. Hay, barns, and cigarettes/cigars do not mix. No smoking should be permitted near and around a barn.

Culling Unproductive Animals

Goats are prolific animals, and a herd can grow very quickly. Culling, or selling for slaughter, will, at some time, be a part of your management if your goats are reproducing.

SELECTING ANIMALS FOR CULLING

- Any animals that are unhealthy, have feet or leg issues, or have problems getting bred are candidates. If they are producing milk and strong enough to stay in the herd, keep them in production through the lactation or sell when the need arises.
- Decide which animals to cull at least a month before breeding season begins. There is no need to breed an animal you plan to cull.
- Use a section of your barn log for listing cull animals.
- If there aren't any strong candidates for culling and you need to reduce your herd, use the same selection process discussed in Chapter 3, Breeds and Selection. As well as physical and milk production attributes, consider attitude, longevity, friendliness, and overall stamina. I assess the entire herd midsummer and give grades A+ through D to all the goats. These grades represent an overall score for many aspects and are simply a reference tool when I need to cull.
- Determine how you will cull. Taking goats to an auction does not guarantee that the animals will be slaughtered. If slaughtering is your preference, find a local slaughterhouse to buy the animals, have them slaughtered, keep or sell the meat, or slaughter the goat yourself. Goat meat is quite popular in our area and can be hard to find. Develop a relationship with a few key people to ease the burden of culling your goats.

Allow visitors to enter the barn only when it is occupied by an owner.

Mixing Goats with Other Livestock

Adding goats to your already established farm animals requires some considerations. Mostly, goats adapt easily to sharing space with other animals.

Horses and goats are good companions, often sharing a stall and paddock area. Be knowledgeable about individual personalities and gradually allow them time together. The sheer size difference can cause problems if they clash.

Cattle and goats intermix well, although they do share a few parasites, such as brown stomach worm, *Ostertagia ssp.* Pasturing calves and goats can be beneficial to both species with proper pasture rotation. Calves will eat clover and other plant species that goats do not favor, and vice versa. Thus the pasture area, with management, will thrive with even plant harvesting by the animals. I've even heard about increased growth rates for both calves and goats sharing a pasture area due to the increased competition. Be careful to monitor pasture growth, moving animals when plants are foraged down to 4 inches.

Sheep and goats do mix, but use caution, since they do share many gastrointestinal parasites, making preventative measures necessary. These include monitoring for parasite burdens, proper rotation of pasture/browse area, appropriate stocking rate for your land base, and cleanliness of pens/paddock area. Also, goats and sheep fight differently, with the goat often taking the beating. While goats rise up on their back legs to head butt, sheep move forward, giving a full force abdominal blow to the goat's exposed belly. Abdominal injuries from blunt force trauma and lacerations can result. Again, mix the species with attention to individual personalities and prepare to make changes to the pen if you suspect or observe problems.

Pigs should be raised separately from goats.

Pigs and goats move quite differently and seem to scare one another, so they are best raised separately. You can, however, rotate pigs into a pasture area after goats, and feed them whey from the cheese making. Pseudorabies, a viral disease of swine, can be passed to goats. The occurrence of this is rare, but be aware.

Poultry generally does well kept with goats. Chicken, guinea fowl, ducks, and geese can be good companions as these birds eat insect larvae and ticks. Coccidia of these animals are species specific. Salmonella, though, is a shared organism, so discourage overhead roosting where the goats are housed, fed, and watered.

Llamas and donkeys make good companions and are used as guardian animals for goats. Llamas do share some parasites with goats, so proper monitoring is important. While these species do a fine job protecting goats, be sure to protect the guardians with routine vaccinations and health maintenance. Keep their hooves trimmed regularly.

Dogs and goats can mix quite well, as in the case of livestock guardian dogs. The two Maremma crosses I've had as guardian dogs have been a pleasure to have with the goats. While the herd protection is instinctual, training, to a certain degree, is needed. These dogs need to bond to their herd, not to humans. This point is *very important*, and everyone who works with your goats needs to follow the protocol to properly establish a guardian dog in your goat herd.

Kidding can be a problem, as dogs are dogs and love to lick bloody things. Mature goats usually keep the dog in line, but a weak doe that

Poultry are good companions for goats.

has a retained placenta may need some time in a health pen, away from the dog. Neighboring dogs can prey on goats, especially if the dogs are permitted to run in a pack. Be wary of such animals and address the concern immediately. Herding dogs, such as border collies, can assist the movement of goats with proper and persistent training. Goats do not move like sheep or cattle and can be frustrating to the dogs. I've seen torn udders, rear legs, and teats from herding dogs nipping at goats as they run in the "wrong" direction.

Developing these management skills takes time. Mistakes will be made and lessons learned. Work with a mentor, keep an informative barn log, and attend hands-on workshops. These are all good ways to improve your ability to care for your dairy goats.

Livestock guardian dogs bond well with goats.

Schedule of Tasks for Seasonable Dairy Goat Operation in Northeast United States

Month	Tasks	
January	• Finish last year's bookkeeping, set plans and budgets for current year • Check animal supplies and order as needed • Visit other farmers and attend workshops	• Check udders on dry goats, trim hooves, evaluate animals, provide exercise • Determine the need for new bucks in the fall
February	• Determine how you will sell this year's kids • Vaccinate goats • Keep feed levels adequate for late pregnancy	• Prepare birthing area and kid pen • Ready kidding supplies
March	• Record kidding information and health difficulties • Care for does and kids • CAE prevention • Keep kid pens clean	• Check hay supply • Make farmer market plans • Find new outlets for your product • Try a new cheese, if desired
April	• Sell all extra kids • Make a pasture plan and buy fencing supplies • Vaccinate remaining kids • Manage internal and external parasites	• Watch for respiratory illness with damp spring weather • Feed first-cut hay before turning goats onto spring pasture • Make soap
May	• Clean and store kidding supplies • Wean kids • Register kids with ADGA and plan for summer shows • Feed goats for peak production	• Take a breath and enjoy the greening of a hillside • Use composted goat manure on your gardens • Trim hooves • Train summer help
June	• Rotate pastures • Monitor parasite levels • Reduce fly population, keep pens clean and dry • Secure hay	• Make strawberry goat milk ice cream *Note:* Milk production is steady now, and farmers markets are in full swing.

Month	Tasks	
July	• Monitor bulk tank and refrigerator temperatures • Clean compressor units for efficient cooling • Don't neglect bucks and kids	• Cull unproductive goats • Compare budgeted and real numbers from first half of year and make adjustments • Take goat cheese salad to a picnic
August	• Provide shade, clean water, and a barn fan during the heat • Watch out for moldy grain with humid days • Rotate pastures carefully—don't over-browse	• Flush buck for fall breeding • Dream of your January vacation • Trim hooves *Note:* Shows and fairs are winding up now.
September	• Keep does in good condition, feeding carefully for milk production and early pregnancy • Breed for February kidding • Monitor for parasites	• Send milk to lab for mastitis/bacteria analysis • Secure second-cut hay • Watch for early frost and deadly wilted cherry leaves
October	• Continue with breeding • Keep goats in good condition—this is important now; fat goats equal birthing problems	• Prepare for winter (fix windows, pipes, hay feeders, waterers) *Note:* Milk production will slowly decrease. This is the last quarter to meet your income goal.
November	• Breed spring kids (mature herd should be bred) • Cull problem animals • Consider attending a winter farmers market as summer markets end	• Think of business and animal care improvements • Cook a decent meal and reconnect with family and friends • Trim hooves
December	• Finish the milking season • Dry herd off according to plan • Watch udders as they resorb milk • Keep animals rotating through milking area at least a few times/week	• Monitor feed program as goats stop milking and go into late pregnancy • Reflect on the past year • Rest!

Chapter 5

Business

We all come into this discussion with different backgrounds, attitudes, skills, and goals. Financial means is one of the considerations discussed in Chapter 1, Getting Started. If having dairy goats puts a financial strain on your family, please reconsider your goals. Dairy goats and their milk should enrich your life. Look at how much money you want or need your dairy goats to contribute to your family income.

Many families are looking at alternative means of income generation, as the world economy falters here and there. In addition to income, food security—both in terms of food safety and availability—adds to people's desire to keep dairy goats. A few goats to simply provide milk and milk products for your family and friends can be kept with minimal expense. Once a goat house and yard are built, general supplies purchased, and animals selected, the main monthly cost is feed.

Let's go back to answer the question of how much money you need from your goats. Determine this figure and work backwards.

It's no small decision to go into the goat business.

Here's a simple example: You want your four goats to earn enough income to pay for the expenses of feeding and taking care of them. Say feed costs (grain and hay) are $8.80/day + supplies $2.20 for a total of $11/day. Labor is the tricky part. A backyard enterprise is not usually paying someone to tend the animals. Paying $8/hour for 2 hours of daily work adds $16/day, and we are up to $27 daily expense.

Four goats giving three quarts daily makes three gallons of milk. Purchasing the milk would cost $12/gallon to equal $36 of *in kind* income. This gives you a net of +$9 a day. Savor that thought for a moment. Your four goats can produce nutritious milk for you with an *in kind*, positive income. You will not be generating real dollars, and your return on initial startup costs may never be realized, but you will have a fresh supply of wholesome milk for you and your family/friends. That is

how your small herd of dairy goats can "earn" you money. Share the labor and expense with another family, and keeping a few dairy goats can be financially feasible.

You can realize real dollars with other income from the goats, such as selling breeding stock, kids, and cull animals; making and selling goat milk lotion and soap; setting up a milk demonstration/petting program and take to institutions/schools/fairs; selling milk for pet food; or selling composted manure. Understand that these added products have their own associated expenses, as well as additional labor. Look at all the figures realistically and keep a financial log. Selling milk and milk products for human consumption requires following local, state, and federal regulations. Some states have exemptions for micro-dairies. Contact your local and state agricultural officials to find regulations in your area.

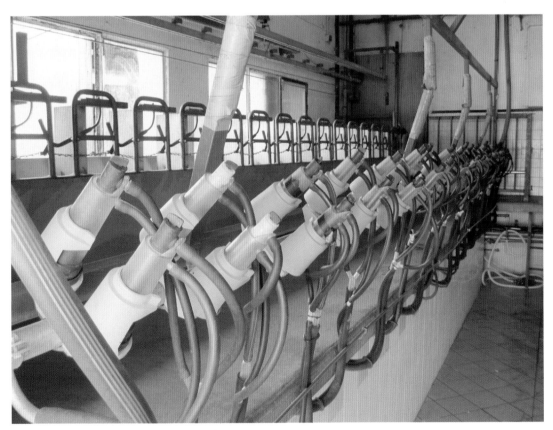
Commercial milking facility setup can be challenging.

Per month ($)	Wholesale Milk[1]	Direct Raw Milk[2]	Cheese-Soft[3]	Cheese-Aged[4]
Income	2,630	6,600	10,340	10,010
Expense:				
Feed – Grain[5]	540	540	540	540
Feed – Hay[6]	500	500	500	500
Feed – Salt/minerals	60	60	60	60
Supplies – Milkroom	70	70	70	70
Supplies – Livestock/bedding	125	125	125	125
Veterinary/medications	55	55	55	55
Milk testing	20	40	40	40
Fencing	30	30	30	30
Building maintenance	40	40	40	40
Fuel/electricity	80	90	100	90
Insurance	80	180	180	180
Cheese supplies[7]	—	—	50	40
Advertising/marketing, packaging, office supplies	—	225	350	350
Dues, travel, etc.	50	150	250	250
TOTAL EXPENSE:	1,650	2,105	2,390	2,370
LABOR*	!	!!	!!!	!!!!
GROSS PROFIT	980	4,495	7,950	7,640
Startup costs/loan payments*	!	!!	!!!	!!!!

[1] Wholesale milk, price $0.55/#

[2] Direct sale raw milk (must meet high quality requirements), price $12/gal

[3] Soft cheese production, price $18.80/#, 1 gal for 1 pound cheese

[4] Aged cheese production, price $26.00/#, 1 gal milk for .7 pound cheese

[5] Bulk purchase, $24/100#, goats eating 3-4#/day

[6] Hay $6/bale, 2-3 bales/day

[7] Fresh cheese additional ingredients (herbs, fruits, etc.)

* ! - Low, !! - medium, !!!- medium high, !!!!-high

Farmers markets offer a good sales opportunity for licensed goat products.

Setting Up Commercially

A commercial dairy goat business requires a higher initial cost. Setting up a milking facility to meet today's standards and regulations can be challenging. Low-interest loans are available through the USDA, including specific programs for beginner farmers. Established farmers also have opportunities through many state and federal grant programs to expand or improve businesses.

Regardless, the single most important aspect of running a successful dairy goat business is your attitude. Set a goal, determine its feasibility, chart a path, and be prepared to wiggle when obstacles come your way. Take risks and opportunities. Both will challenge and allow you to grow. Be practical and understand your limitations. Communicate with others joining the goat-keeping team, and learn from more experienced farmers. Explore your potential market and customer base before spending money on goats or a facility.

There are many good resources for setting up budgets and break-even calculations for commercial goat dairies. An excellent resource is *A Guide to Starting a Commercial Goat Dairy*, by Carol Delaney, which is easy to follow and full of useful information. See also www.uvm.edu/sustainableagriculture, as well as other resources in Appendix 2.

Let's look at a monthly budget scenario for a twenty-goat dairy. These numbers given are estimates for planning purposes only. Create your own chart using current figures and other expenses for your locality.

MONTHLY BUDGET

(Twenty goats average production 1 gal/ day, less 50 gallons for farm use)

The table at left contains general information, not taking into account some very real expenses, such as labor, taxes, and startup costs. Selling milk wholesale does not have the labor expense for a value-added product or associated startup costs. Direct raw milk sales need to be legal in your area and also have a customer base to support the sales. Strict attention to bacteria counts, animal health, and cleanliness are necessary. Soft goat cheese requires startup equipment and a customer base. Many markets are saturated with fresh goat cheese and are not looking for new producers. Aged cheeses require startup equipment, a cheese aging facility, and a customer base. Be prepared for a delay in cash flow with aged cheese, as there is a waiting period for income as the cheese matures.

Many areas of the United States are experiencing a growth in sales of local agricultural products. Goat cheese is a popular local item sold in farmers markets, local food stores, farm stands, and through farm shares.

Local restaurants, caterers, and private institutions are also looking for area cheeses. Develop relationships in a variety of outlets. If you have a rainy season and farmers market sales are down, reach out to local chefs to include your cheese on their menu. Just remember that being local does not guarantee a sale. High quality and product consistency are important, as are sound business practices. Believe in and be proud of your product. Set a balanced price. Don't oversell inventory, leaving dedicated customers with empty shelves. Use social media to get the word out: cute goats can help create demand. Marketing strategies can be beneficial. Crafting a high-quality product is absolutely necessary.

As mentioned, sale of breeding stock can be a good way to increase real dollars earned. Breeding stock needs to come from superior bloodlines, with excellent production and body type. In addition to having linear appraisals, many goat keepers show their animals at sanctioned goat shows. Showing goats has plenty of benefits. Financially, winning animals are worth more with the sale of breeding stock. Shows and the time preparing for them create

Showing goats is beneficial in many ways.

a wholesome environment for family bonding and teaching children responsibility. Children make friends from beyond their neighborhood and school, and learn to respect others.

Breeders learn from the judges about improving their specific animals and goats in general. Goat shows generate interest from the general public, and improve the overall attitude toward goats.

There are difficulties to overcome with showing. Many long days turn into long drives home. Goats have to be transported safely to and from the shows, and often be on display at fairs for several days. Showing is real commitment both in terms of time and money. Proper grooming and training the goats to behave in the ring is paramount to achieve at shows. Goats need a health certificate to travel outside most states. If you suspect an unhealthy animal at a show, speak to the organizer. Take care to reintroduce show animals to your general herd, understanding the need to possibly quarantine.

Expenses can be recouped with the sale of animals. Showing goats teaches participants how to gracefully accept disappointment and hopefully learn from it. Judges give reasons why an animal is at the front of the line. Listen and learn so improvements can be made.

If you have interest in showing, spend time with a goat breeder helping to groom and train the goats. Be helpful and considerate of their

Above: Standing for the goat judge is probably more harrowing for the farmer than the goat. **Top:** Show ribbons are a fun and simple reward for raising an excellent herd, and may increase dairy sales opportunities for your family or business.

time. Realize the commitment of the show circuit before investing money in traveling equipment and animals. Encourage children to participate by having them help show other people's animals. Teach others to show your winning animals, sharing the joy and excitement of achievement. Volunteer to help run or financially support a show in your area.

Just as goats are adaptable and resourceful, you can discover additional avenues of income with your animals. In addition to milk and dairy products, explore possibilities with animal products such as meat, hides, crafts, and manure.

A successful business requires work, dedication, planning, and knowledge. Luck is really cleverly disguised toil. Remember your wild cards, and play them carefully. Look at risks with a determined understanding and an eye on your goals. Adapt and make changes to keep a healthy life balance.

Top: Tanned hides can be sold for additional income. **Above:** Goats await their time in the ring. **Right page:** Goats filing out to summer evening pasture—it's dinner time! *Danielle Mulcahy*

Chapter 6

Breeding

Dairy goats mate and then give birth to begin producing milk. In the wild, animals easily handle these aspects of life. With the domestication of goats, we in part have taken the reins and now participate in planning these activities. Let's start with the basics.

Goats are mammals. They produce milk after giving birth. Mating or breeding is the first step in goat reproduction. A doe ovulates, producing eggs, which are fertilized in her oviducts with a buck's sperm. The period of ovulation is called estrus, or heat. The estrous cycle lasts 21 days (more on this later). The resulting pregnancy, or gestation, is basically 5 months, with a range of 145 to 155 days. Parturition means giving birth. A goat generally gives birth to twins, with singles and triplets also being common.

While the actual mating is generally simple, the Who, Why, Where, and When of breeding requires Work.

Buck Selection

This is the Who and Why. Who will make a good partner for your doe? Why select a particular buck? To answer these questions, look at your established goals. Are you looking for a healthy buck to easily breed your mature doe so you have family milk for next year? Are you raising purebred lines to show and sell breeding stock? Do you want to improve specific body traits or milk components of your herd? These types of questions answer the Who and Why. Consider this a long-term project, like sewing that heirloom quilt or playing a winter-long game of RISK. With

eyes on your goals and attention to details, matching genetics to achieve the desired results has great rewards.

Breeding to a particular animal line to improve specific traits requires persistent homework. Attending fairs and shows (in the ring or sideline), talking to goat farmers and breeders, joining goat breed associations, and studying pedigrees are ways to learn about dairy goat type. A linear appraisal assesses your animal's type, identifying strengths and weaknesses. Genetics are literally a shot in the dark, but there are tools that shed light on the subject. Production type index (PTI), predicted transmitting ability (PTA), and estimated transmitting ability (ETA) are values used to make breeding improvements. In selecting a buck, pay close attention to the performance of his dam (mother), sisters, and daughters to see the heritability of traits.

Genetic Values to Consider
Predicted Transmitted Ability (PTA)
This number indicates the genetic potential that is expected to be passed onto offspring for a given trait (production and type). These values are reached through complex calculations of statistical data, including DHI production information, linear appraisal type scores, and parent and offspring evaluations. The number is the potential halved, indicating what each parent could give to an offspring.

Estimated Transmitting Ability (ETA)
This number expresses a young buck's genetic potential of production and type before he has daughters of age to contribute statistical data.

Linear Appraisal and Genetic Values

Linear Appraisal – Linear appraisal is a systematic evaluation of individual type traits that affect a dairy goat's production, health, and longevity. A trained appraiser evaluates animals for functional conformation based on established traits. These traits must have economic value, be moderately heritable, and be possible to assign a value with repeatability. Thirteen primary traits and one secondary trait are used. These traits range from dairyness and strength to rear legs and udder shape. Organized goat groups will often plan an appraisal event to share the expense of hiring an appraiser.

Production Type Index (PTI)—This number combines both DHI production data and linear appraisal type scores to express economic merit of an animal. There are two indexes, PTI 21 and PTI 12. PTI 21 weighs *production* twice as much as type, while PTI 12 weighs *type* twice as much as production. Decide which you want to try to improve more and choose the appropriate higher score. Of course ideally you want to improve everything at the same time—like NOW!—but this is genetics and predictions, not promises. Remember you are working on that heirloom quilt. This index is relative to breed averages, so be sure to compare, for instance, Alpine to Alpine, not Alpine to Oberhasli.

Genetics is truly a fascinating subject. If this interests you, take time to learn about specific traits, their heritability, and how to track offspring development. A lot can be learned, and specific lines of goats can be raised with a trained eye and some homework.

Records from a DHI program can help goat owners track breeding progress in terms of milk production and quality. Creating individual animal records for daily or monthly production, milk protein and fat percentages, somatic cell counts, and other specific goat information is helpful when evaluating your goats to help set breeding goals. We'll look more into these specific ideas in Chapter 10, Milk and Milking.

If you're reading this saying I don't want to do all that work, it's okay. Take a breath. Having a couple of healthy goats in your backyard for family milk and fun doesn't require this. Just make sure your commitment and resources match your goals and are well communicated to involved family and friends.

Finding the right buck for your breeding program can take some time, even if your program is to simply have a pregnant doe. Often, backyard goat owners prefer to not own a buck. Plan in advance and locate a buck to breed your doe before she is in heat. County fairs or your local agricultural supply store are good places to ask about local goat farms. If you want to buy a buck from pedigreed stock, you may have to plan a year in advance. Buying a specific buckling in the spring could mean making a deposit in early winter, before he is born. Artificial insemination (AI) is another option.

So many goat breeds, so little time! Each breed offers unique personalities and traits. *Shutterstock*

Line Breeding and Crossbreeding

Pure breeding, crossbreeding, inbreeding, and line breeding; do you want the genes straight up, shaken, or stirred?

Following genetic lines and selecting desirable traits require keen observation and a steady goal. Develop a system to keep track of your breedings, such as a spreadsheet and pedigree diagrams. Specific breedings can yield superior offspring. The downside of concentrating genes is that negative traits can be expressed as well. *Cull* is an important word when selectively breeding goats or raising goats in general. As a longtime award-winning goat breeder told me, "The only way to improve your herd is to adopt the word CULL in your management." Culling improves the herd and also your bottom line. And please, don't sell bad traits to others.

Pure breeding – A purebred animal is offspring from parents of the same breed. Traits characteristic to that particular breed are strengthened. The benefit of pure breeding is a higher predictability of production and type.

Cross breeding – Crossbreds are offspring from parents of different breeds. This line of breeding brings an initial boost in many traits, the hybrid vigor. This vigor is not generally passed onto the next generation, thus cross breeding can be the end of the line, in terms of genetic improvement.

Inbreeding – An inbred is offspring from parents that are closely related. Some define this as related to the first degree, as in offspring to parents or siblings. Yes, there can be a concentration of good traits with inbreeding, but it also risks the concentration of negative traits. A good understanding of genetics and breeding, and a strict program of culling, is necessary to improve a herd using occasional inbreeding. Again, the word is cull—selling direct to slaughter and not selling inferior animals as pets.

Line breeding – A line-bred goat is the offspring from parents that are related more distantly than to the first degree, such as the granddam on the sire's side being the doe's dam. This is a concentration of the granddam's genes stirred with other genetic material. Once again, studying pedigrees, making careful observations and appraisals of the offspring, and maintaining a strict cull program can result in a strong and predictable line of goats.

Purebred animals are the gold standard in multiple generations. *Shutterstock*

Artificial Insemination

AI is a breeding technique that fertilizes a doe using previously collected and stored semen. The goat breeder can specifically select desired genetics without owning a particular buck. AI allows for greater genetic diversity and can improve both production and type in less time (fewer generations).

The semen from bucks with superior and proven genetics is collected and processed and then packaged in small straws. Most often, the straws are stored frozen, in liquid nitrogen. Occasionally, fresh semen will be used for AI.

The downside of AI is that the necessary equipment and semen can be expensive. Finding an experienced technician or learning the procedure yourself can be difficult.

Hormone shots can be used to coordinate the breedings in your herds or those of nearby herds if a technician is being hired.

The basic steps are explained below. *Warning:* advanced knowledge and experience are necessary to perform this technique. AI for goats is available in many parts of the country. Ask a goat mentor or veterinarian for advice if you would like to use this procedure for breeding your doe.

Cleanliness is important for this technique, both for the doe's health and the viability of the sperm. To use, thaw the selected straw with care to not harm the sperm. Load the straw into an insemination gun and ready it for insertion into the doe.

The receiving doe has to be in full estrus or heat. She needs to be safely and securely restrained in a breeding or milking stand with aid to hold and calm her. The goat may be nervous and jump around.

Place an AI light into a clean, well-lubricated speculum and carefully insert the speculum into the doe's vulva and vagina. The reddish-purple cervix will lie ahead. A doe in full heat will have a white mucous coating on her cervix. Center the speculum at the vaginal end of the cervix and insert the insemination gun into the speculum. Maneuver carefully to avoid contamination or injury to the doe. Continue inserting the gun through the cervical canal. Stop at the uterine end of the cervix.

Slowly deposit the semen at the uterine end of the cervix or *slightly* inside the uterus. The semen dispersal should take 5 or 6 seconds. Remove the insemination gun without releasing the plunger. Carefully withdraw the speculum from the goat.

Record the breeding information in your barn log and on the individual goat sheet. The semen straw can be taped to the barn log for future reference. Note to watch the doe 3 weeks from the insemination date for a repeat heat and 5 months for a possible due date.

Clean up your breeding area and wash and sanitize all reusable equipment.

Note: Use caution when breeding with naturally hornless, polled goats.

Breeding Polled Goats

Be careful breeding two polled (naturally hornless) goats, as an intersex (a sterile animal showing both male and female characteristics) may result.

Doe	Homozygous buck	
	P	P
P	PP Polled Intersex if female	PP Polled Intersex if female
p	Pp Polled Fertile	Pp Polled Fertile

Doe	Heterozygous buck	
	P	p
P	PP Polled Intersex if female	Pp Polled Fertile
p	Pp Polled Fertile	pp Horned Fertile

Note: Polled Doe is always heterozygous. *Adapted from Goat Medicine, 2nd Edition. Mary C. Smith, David M. Sherman, Wiley-Blackwell, 2009*

Use a strong healthy buck for breeding.

Location

Keeping a buck requires space, feed, time, and the commitment to care. If you can't provide these, consider taking your doe to a local farm for breeding. If you take your doe to a farm for stud service, be sure the herd and bucks are healthy. Communicate with the owner regarding breeding fees, including a second breeding if needed, and registration of the buck if you plan to register offspring. A service memo is needed from the buck's owner for registering offspring. Get this before leaving the breeding farm and keep it in a secure location. *Only take healthy does to another farm for breeding.*

Bucks brought to your farm should be quarantined for a minimum of 30 days—I recommend having the buck for at least 90 days prior to breeding. This gives him time to adjust and be at the top of his game! Homegrown bucks also need to be in prime condition. Keep the males in good shape throughout the summer, with plenty of fresh water, browse and feed, salt and minerals, and a good internal and external parasite-control program.

Regardless of where the buck is from, you want a sound, healthy buck breeding your doe.

Timing

When do you breed your goat? Gestation is 5 months. Think ahead and project for specific reasons. What are the priorities for the upcoming year in connection with your goals? Wanting to sell kids for a seasonal holiday like Easter? Figure out the expected growth rate for kids to reach desired market weight and plan back accordingly. If you're in a wedding party in March, 2,000 miles from home, simply breed in November for the goats to give birth in April.

I breed our goats in mid-September to start birthing in mid-February, for several reasons. First, kids born in colder weather are historically healthier on our farm. Second, since we market a lot of our goat cheese through farmers markets, the full-time task of parturition and kid rearing is minimal by the start of market season.

Many breeds of dairy goats are seasonal breeders, meaning the ovulation hormone cycles lie dormant for many months of the year. The increased darkness of late summer and autumn (August through December) jump starts these hormone cycles to begin the breeding season. These goats are known as short day breeders. Other breeds, mainly from tropical roots, breed and give birth year round.

Corresponding to this, the buck begins his rut in late summer. While you may be uncertain whether your doe is in heat, there is no question about the buck's rut. The musky odor lingers in the humid air of late summer. Bucks begin urinating on their heads and legs and sometimes develop an attitude. Breeding season is here!

Obsessive tail wagging, bleating, "jumping" other goats, redness and swelling of the vulva, a clear vaginal discharge, and a slight drop in milk production are signs of heat. Walking a buck *outside* the pen of a questionable doe will give you the answer. Does are generally in heat for 36 to 48 hours. The heat or estrous cycle averages 21 days, with a range of 17 to 24 days. Note on your barn calendar the first time you see your goat in heat. Calculate 3 weeks from this date and watch for a repeat heat at that time. Put the doe and buck together according to your planned "When." If taking your doe to another farm, notify the breeder in advance with a FYI probable date, preferably after you observe a couple of estrous cycles.

Watching for goats in heat/estrus is part of the daily observation, especially August through January.

While they may not stand to be bred in the early part of the heat, try breeding again 8 to 12 hours later. Does seem to be most receptive in the 18th to 36th hour of the actual heat. Ovulation occurs 24 to 36 hours after the initial stages of heat.

Estrous Cycle

The female goat estrous cycle is defined as the period between two estrus (heat) dates. The cycle runs an average of 21 days, with a general range of 17–24 days. There are four main phases of the cycle: Proestrus, Estrus, Metestrus, and Diestrus.

Estrus is the physically active phase of the cycle, lasting 24 to 48 hours. The maturing follicle releases estrogen, the primary female sex hormone. Tail wagging, loud cries, rubbing or jumping other goats, clear vaginal discharge, and swollen vulva are signs that the doe is in the estrus or heat phase. The goat is receptive to breeding now, generally in the mid to late period of this phase (12–24 hours after the onset).

Metestrus is the phase when ovulation occurs. The developed follicles (6 to 9 mm in diameter) contain a mature oocyte. The oocytes are released during follicular rupture, generally 24 to 36 hours after estrus begins. Luteinizing hormone (LH), released from the pituitary gland, matures the follicle to ovulation. The released oocyte travels to the oviduct to be fertilized by the deposited sperm. The ruptured follicle is altered to a new structure called a corpus luteum (yellow body) or CL. The CL,

which is creamy-yellow and about 8 mm in diameter, secretes progesterone.

Diestrus is the longest phase of the estrous cycle. The period lasts 15 to 19 days. The CL continues secreting progesterone, waiting for a hormone signal to indicate if the released egg was successfully fertilized. If so, the CL remains throughout the pregnancy, producing progesterone and halting the estrous cycle. If not, the CL will stop producing progesterone. Prostaglandin, a hormone secreted by the uterus, will push the CL to regress, causing follicles to begin development.

Proestrus is the period of follicular development. The phase lasts 2 to 3 days. Follicle stimulating hormone (FSH), released from the pituitary gland, directs the maturation of the ovarian follicles. Recent research indicates that follicles develop in a wave-like pattern, with three to four waves being prevalent per cycle. The goat is not sexually receptive at this time.

And the cycle begins again.

Understanding the estrous cycle of your goats will go a long way toward happy and healthy husbandry and herd management.
Shutterstock

Bucks and does go through a mating ritual, including rubbing on each other, tail wagging, sniffing, and squatting. Bucks will often snort and flash the flehmen response, curling their upper lips to help float pheromones through their nostrils. Depending on many factors, the buck may mount the standing doe and quickly mate, ejaculating semen as he throws his head and body forward and up. Other times, they may circle and play for many minutes, with the buck making attempts but not achieving ejaculation. Upon complete ejaculation, semen may be seen on the vulva of the doe. The buck often goes for a second breeding after a brief rest.

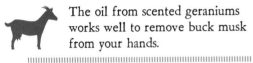

The oil from scented geraniums works well to remove buck musk from your hands.

Note the mating date on your calendar and project 5 months forward for a possible kidding date. Also mark 3 and 6 weeks ahead on your calendar to watch for repeat heats. If you don't see a repeated heat in this time period, your doe is probably bred.

The simplest scenario, if you have a buck, is to put the goats in the same pen around the date you determine for breeding. You may not see the mating and will not have a firm due date. Watch for heats or test for pregnancy (discussed later) to confirm a successful breeding. If you do not have a buck, you will need to spend some time with your goats observing heats and figuring out when to breed them. If you are renting a buck for the breeding season, I highly recommend learning the buck's health status and following the 30-day quarantine. You don't want to bring health problems, such as respiratory infections or resistant parasites, to your does.

Infertility (not getting pregnant) may occur in goats. Causes can be nutritional deficiencies, uterine infection, hormonal imbalances, or reproductive tract defects. Consult your veterinarian or an experienced goat breeder to help identify the cause and ways to correct the problem. You may have to hold this doe to the next breeding season or use artificial hormones to jumpstart the estrous cycle.

A small (2 to 5) percentage of pregnancies result in naturally occurring abortions. If you

A strong buck will ensure strong kids. Mark your calendar—kids are due in just 5 months! *Shutterstock*

experience a large percentage of abortions, consult your veterinarian to determine if your animals have been exposed to abortion-causing organisms, such as toxoplasmosis, listeriosis, Q fever, or chlamydiosis.

A goat pregnancy can be confirmed through a blood test or ultrasound. The blood test measures the presence of pregnancy-specific Protein B (PSPB). This protein is only produced by the placenta of a growing fetus. BioPRYN is accurate and cost effective. Blood samples should be drawn by a trained person. For more information, try www.biotracking.com.

Separate young doelings and bucklings by 3 months of age. Bucklings can be fertile at this time. Be sure young does have adequate weight before breeding. A sound breeding weight is 75 to 80 pounds for standard-sized dairy goat breeds. With proper care and feeding, this weight can be reached by 7 to 9 months, with births occurring when the does are around one year old.

Pregnancy (Gestation)

Goats should stop milking around 8 weeks prior to kidding (3 months into the 5-month gestation). This rest allows the doe to keep more nutrients for herself and to grow the fetuses. With twins or triplets weighing 6 to 10 pounds each at birth (for full-sized goats), the fetal demand is incredible. Also, as the fetuses come to full term, the doe's body capacity for ingesting feed is minimized. This can lead to metabolism issues such as pregnancy toxemia (ketosis) and other preparturient health concerns. Exercise, fresh air,

proper body condition during pregnancy, minimal parasite load, and a sound feeding plan are important to the health of a pregnant doe. Read more about this in Chapter 3, Feeding, and Chapter 9, Health Care.

That is one healthy, happy, super-pregnant doe. *Shutterstock*

Rest is as important as play for newborn kids. *Danielle Mulcahy*

Chapter 7

Birthing
(Kidding or Parturition)

Your dairy goat giving birth to kids will be exciting, scary, heart-warming, and heart-wrenching all at the same time. After being present for literally hundreds of birthings, my palms still get sweaty and my heart skips some beats. Telling yourself to stay calm and actually doing so can be challenging.

Pay attention to your five Ps (Prior Planning Prevents Poor Performance) to help make the birthing process go more smoothly.

One month prior – Find your birthing and kid feeding boxes. See the kit lists in Appendix 1. Check for supplies and order or buy items as needed. Also determine your needs for pens,

Kids like a secure area to creep away from adults. *Jay Iversen*

CAE prevention, and sale of offspring.

Two weeks prior – Prep your kidding area and pens. Birthing pens should be at least 4 feet x 4 feet (1.5 meter square). Use an established area or construct something for this purpose. Avoid sharp metal edges, loose wires, and widely spaced boards. Newborn kids do a lot of wiggling around, so watch for areas that can trap them. Use sturdy and secure construction materials, as even very pregnant goats are quite active. Make sure your pen is draft free, dry, and well bedded. Straw and wood shavings are common bedding materials. You can stack hay bales around the exterior of the pen to add an insulation layer in cold weather. I don't recommend using heat lamps. Create a small creep area where kids can go to hide.

If you are separating the kids as soon as they are born, have a pen or box ready for them. This pen also needs to be draft free, dry, and well bedded. I use large plastic dog crates. They are easily scrubbed and sanitized, lightweight and mobile, and securely hold kids. Kids are kept in these for 1 to 3 days, with time to run around during feedings.

One week prior – Trim your fingernails! Have everything ready to go. You don't want to scramble to find supplies or clean a pen when the doe is birthing. Be prepared for the task at hand.

Birthing

While does don't cry, "My water broke—grab the bag and let's go!" they do give you signs of labor. Your job as goat keeper is to observe the signs and learn this kidding language. While there is basic structure to the birthing process, every doe goes through it her own way. Here's where it can get tricky. Learn by assisting your goat mentor with a few births before tackling this on your own.

THE BASIC PRE-BIRTHING PROCESS

One to 2 weeks before kidding, the udder begins filling with milk. The fetuses "settle" or drop into a lower position in the doe's abdomen. Remember there are usually two and sometimes

Spinal ligaments soften and almost disappear near birthing.

three kids inside. The pelvic ligaments that lie on either side of the spine soften in preparation for dilation of the birth canal.

Two to 4 days before kidding, a thick, creamy mucous plug can be observed seeping from the vulva. Spinal ligaments continue to relax, particularly in the area by the tail head. The softened ligaments between the tail head and pin bones create a "hollow" that is soft and jelly-like. The doe's behavior alters some, and she spends time away from the herd. She shows signs of being uncomfortable, often groaning and shifting positions. Be sure the doe is eating and drinking, particularly if she is a timid or over-conditioned goat. Pregnancy toxemia (ketosis) can kick in now—be observant and ready to act.

One to 2 days before kidding, the udder gets taut, and the doe becomes more vocal and restless.

On the day—or night!—of kidding, muscles and ligaments around tail head "disappear."

The udder is quite full, possibly leaking some milk. Vaginal discharge of thick and translucent mucous is noticeable. The vulva swells and becomes placid. The doe may paw at ground, making a nest.

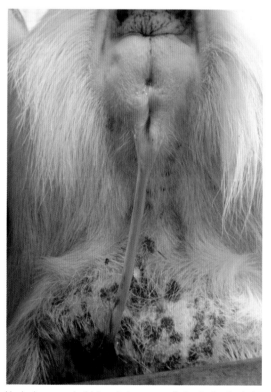

Not all does show these signs and not always in the timeframe given. Birthing is truly a special, magical occasion. Don't get so stressed that you miss this wonder. A doe will pick up on your nervousness. Stay calm and help her relax.

LABOR

Experienced goatherds develop a method of knowing when a doe is ready to kid. Some goats make different sounds. Their behavior changes, and their breathing becomes shorter and faster. For me, it's their eyes. Stare into a laboring doe's eyes and you may discover this clue as well. Their eyes shift focus and zone out. Endorphins and other hormones are in full flow, creating a birthing euphoria. Their bodies are primed for the continuation of life, and there is simply pure magic in those eyes.

Vaginal mucous is thick and clear, generally beginning to drain within 24 hours of birthing.

Goats, as with most domesticated animals, really do fine giving birth on their own 95 percent of the time. We humans often intervene more for our sake then theirs. But that 5 percent when we are needed can mean the difference between life and death, and no, they don't tell us in advance! Have your doe in good body condition before and during the pregnancy. An over-conditioned (fat) doe will have more problems during pregnancy and birthing.

Many people use baby monitors or remote cameras to keep aware of birthings. Others break out the sleeping bags and hunker in the milk stall for the night. Raising the kids with CAE prevention *requires* being present for the birthing, as the kids cannot drink even one sip of milk from the doe. Even though our goats tend to birth during the day, nighttime vigilance is necessary. I make careful observations on my evening check of the barn, set the alarm, and get up during the night—sometimes every 2 hours—to check on does.

Here are the basic steps of active labor (also see the birthing video at vimeo.com/user29715351).

The doe selects a birthing area, oftentimes up against a wall! Clear this area of obstructions, including water buckets.

Quick and shallow breathing (think Lamaze) and frequent contractions indicate active labor has begun. A thin-walled, fluid-filled sac will appear first. This sac helps widen the birth canal. It may break inside the doe with a gush of fluid. The doe may get up and drink this fluid. Kids should appear within an hour.

The sac containing the kid and amniotic fluid will appear next. Normally goat kids are born in a prone float position, presenting the two front feet first with the head positioned between the front legs. They can also be born rear legs first (see note on page 96). After the head crowns through the vulva, the shoulders, hips, and trailing rear legs should come out with a few pushes. If the kid is getting hung up, assist by *gently* holding and pulling the front legs, applying slight pressure to birth the kid. Work with the doe's contractions and angle of delivery.

CAE Prevention

Caprine arthritis and encephalitis (CAE) is a viral infection that is widespread among dairy goats. Most CAE infections, though, remain subclinical (symptoms not expressed), with only around 20 percent of CAE-virus-infected goats actually showing symptoms in their lifetimes. Symptoms in adults include arthritis, swelling of knees, swollen and hard udder that lessens milk productivity, and a chronic pneumonia. Kids can exhibit symptoms of incoordination and eventually paralysis, generally of the rear legs.

The CAE virus is mainly spread through kids ingesting colostrum and milk from CAE-positive does. Some sources report horizontal transmission through direct contact at feed bunks and waterers, or through repeated use of needles and equipment contaminated with blood.

The recommended CAE virus prevention program includes removing kids as soon as they are born, raising kids away from the adults, and feeding colostrum and milk that has been heat-treated to kill the contaminating virus. See the schedule below.

Information on the CAE virus, prevention, treatment, and testing continues to evolve. Discuss with your veterinarian or goat mentor, or follow reliable resources, such as www.merckmanuals.com or the Maryland Sheep and Goat Page, www.sheepandgoat.com.

1. **Prevention Program**

Remove kids from does immediately after birth. Do not allow the doe to lick the kids, or the kids to suckle. Towel the kids dry and place them in a clean, prepared area, away from any animals that are potentially positive for CAE virus.

Feed kids heat-treated colostrum within an hour of birth (see Chapter 8, Kid Care). Continue to feed heat-treated colostrum for 2 to 3 days, and then transition to pasteurized milk or a replacement that is free of CAE virus.

2. **Heat-Treated Colostrum**

Colostrum must be heated and held between 133 and 138°F (56 to 59°C) for 1 hour to deactivate the virus. Use an accurate thermometer. Do not exceed 138°F (59°C). A high-quality thermos, wrapped tightly in warm towels, will generally keep the colostrum at a proper temperature after it is carefully heated on a stove or with a double boiler. Pre-warm the thermos with 138°F (59°C) water, and check the temperature after 1 hour to ensure your setup. Empty the water and pour in warmed colostrum, sealing tightly and wrapping with warm towels. Check the temperature after 1 hour. Successfully treated colostrum can be fed, chilled, and stored in a refrigerator for short-term use (within a day) or chilled and frozen. Be sure to label colostrum.

Prep yourself to help with kidding: wash hands and forearms (remove large rings and other jewelry) with mild soap and water, and dry. Have long-sleeved gloves, lubricant, towels, and iodine dip ready.

Clear the kid's nose and mouth of mucous by wiping with a clean towel. Remove kid from the birthing area if you are raising CAE-free. Otherwise bring the kid to the doe for her to lick. If necessary, assist by drying with clean towels. Rub the rib area briskly to help stimulate the lungs. Kids can be hung upside down and swung gently to clear the nostrils of mucous, or tickle a nostril with a small piece of hay to make the kid sneeze.

The umbilical cord should tear naturally with delivery. Dip the navel in iodine using a small container. If you spill iodine on your own skin, wash immediately.

The doe will often take a break and lick this kid before proceeding with labor. Complete birthing of twins or triplets can take 20 minutes to an hour. Be patient.

Bring the doe a bucket of warm water with blackstrap molasses added (2 to 4 tablespoons of molasses per gallon of water). This treat will recharge the new mom and get her up and moving.

Expulsion of the placentas can begin shortly after the kids are born. Some does will eat or drink fluids from the placenta.

If the kids are going to nurse from the doe, make sure milk is flowing from the udder by expelling some milk from the teats. Help the kids begin to nurse when they are ready.

You may need to assist by holding the rear legs and applying slight pressure as the kid is pushed out. Again, work with the contractions and angle of delivery. The hips, ribs, and shoulders can sometimes get stuck. Move quickly to get the kid out. Once the umbilical cord is broken, the kid will want to take a breath. Air is preferable to amniotic fluid!

Retained Placenta

The placenta is normally pushed out by 4 to 6 hours after kidding. Any membranes that remain attached to the uterus 12 hours after kidding are called a retained placenta. Retained placentas can sometimes occur following a difficult birth or due to nutritional deficiencies. Does can be exhausted after birthing a large kid or having an abnormal delivery. There is no more "umph" to continue contractions to pass the placenta. Sometimes a rest after drinking a bucket of warm molasses water and eating some good hay will be enough to get the doe to pass the placenta. The flow of the hormone oxytocin aids uterine contractions. Oxytocin, which is secreted from the pituitary gland, is released during milking. Thus the kids nursing or your milking of the goat can help dispel any membranes that remain after kidding. Deficiencies of selenium can also cause this condition. Exercise, sound nutrition, and good body condition for birthing all help avoid this problem. Consult with your veterinarian if the doe does not pass the placenta in a couple of days to determine the best treatment.

Do not pull on the membranes. This action can tear the uterine lining. If you have a working dog living with your goats, separate the doe with a retained placenta. You don't want the dog to tug on these membranes. I have wrapped a retained placenta in a plastic bag and tied the bag to the hanging membrane close to the vulva. This "package" keeps the goat from stepping on the membranes and also concentrates the weight to help in releasing the placenta.

A uterine infection can follow an ignored retained placenta. Tend to this situation and correct it as soon as possible using recommended advice from your veterinarian or goat mentor.

As stated, 95 percent of the time kids are birthed with no problems. Strive to be present to support the doe, assist with slight pressure to help a stuck kid, clear mucous from the nostrils and mouth, and dry the kid. If raising the kids CAE-free, you *must* be present to retrieve and remove the kid immediately at birth.

And the 5 percent of the time, problems do occur. Kids should start being born by 1 hour after the initial water sac appears. If the doe is having strong contractions but not progressing with delivery, you may need to help. Call for assistance from an experienced animal keeper, veterinarian, or at least someone to hold the goat and keep you calm!

Remember: Goats do generally give birth without any problems! Have a knowledgeable person at the ready in case there is a difficulty. Always keep OB lubricant in your birthing box; a little lube can make a big difference. Be prepared.

This doe is ready to give birth—note her dazed expression.

Kid being presented in normal position—prone float.

First the head crowns through the vulva.

Then the hips and rear legs follow.

Above: Does begin cleaning newborn kids almost immediately.

Right: Kids can also be born with rear legs coming first. Distinguishing front and rear hooves can be hard with everything else going on. Rear legs will show the sole (bottom) of the hooves. Front legs will come with the top of the hoof showing.

Opposite: Clean hands and a warm towel are often all that's needed immediately after a successful birth. *Danielle Mulcahy*

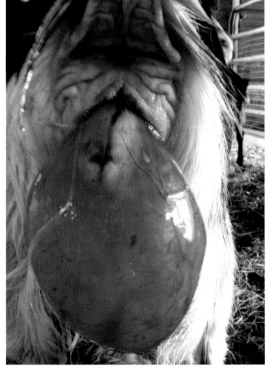

Abnormal Birth Positions

The following paragraphs will make more sense after you have helped with a few births. The first time you help correct an abnormal presentation will probably be nerve wracking. I will go into detail discussing how to correct this first position to help you create a mindset to follow when correcting any abnormal birth position. What's most important is to stay focused on your task and understand your limitations. Get the kids delivered safely without injuring the doe or yourself. When entering the doe's uterus, work as cleanly as possible. I wear long-sleeved gloves to reduce the risk of infection.

Keep a copy of the Correcting Birth Positions Chart from page 102 in your birthing box.

HEAD NOT CONNECTED TO PRESENTING FEET BONES

With multiple fetuses being born, they sometimes crowd the birth canal and get tangled in the chute. If you see two feet and a head but the doe is having problems, make sure both feet belong to the head. Apply lubricant to your long-sleeved glove or clean hand and slide your fingers up over the skull and down the neck. Proceed to the shoulder and then to the front leg. This should connect to the exposed foot. Repeat on the other side of the kid.

If *one* foot does not connect, gently push this leg back into the doe. Go back up to the head and follow the shoulder down to the correct leg. Cup the foot with your fingers to avoid tearing the uterine or birth canal lining, and swing forward. If necessary, push this head back into the uterus to give yourself room to swing the leg forward. Deliver this kid with normal presentation.

If the head does not connect to *both* feet that are out, push that head back in. Find the head that is connected to the feet that are out. Most likely, it will be tilted to one side. Grab the nose or mouth, and turn the head forward. Cradling the skull or jaw in your palm while you turn the head helps to keep the head in place. Pull gently on the legs while you are holding the

Newborn kids need to rest right away. *Danielle Mulcahy*

head and all should slide into a normal position. Continue to aid the doe while she delivers this kid. Once this kid is out and breathing, you can go back inside the doe and bring the other kid out. The feet and legs may have joined the head now that there is more room inside. If you still only feel the head, follow the neck as before until you find the front legs. Again, cup the hoof with your fingers and gently swing the leg forward. Repeat on the other side. The kid should now come out easily.

Points to remember: Work methodically. When making corrections for abnormal positions, you'll need to push the fetus back into the uterus. The doe isn't going to like this and will fight you. She will give in and let you help once you take control of the situation. Slide your fingers and hand along the kid's body to determine what is what. Glance at the doe to help you imagine what you are feeling. If you get confused, stop for a moment and breathe. Begin back at the body part you know and start again to determine what's going on.

Does that have been assisted should be given a preventative course of antibiotics (I recommend Procaine Penicillin G) or other infection-preventing remedies. Take her temperature every 12 hours for 3 or 4 days to monitor for a fever. This rise in body temperature could indicate an infection.

True story: I was helping a ewe deliver years ago. The ewe presented a head and a foot, but was obviously in distress. I put on a sleeve and went in to help. I found another head inside connected to another foot. I tried for a while to find the other foot to go with what I had. Pushed head back in, explored, explored, explored. I reached my limitation and called the vet. She came and explored and determined the lamb had two heads! Yes, the lamb had two heads, three ears, and two front legs. This creature did not make it, but we were able to save the ewe. Incidents such as this are reminders that we don't live in a perfect world.

FEET WITH HEAD BACK
I find this to be one of the hardest positions to correct, because the head wants to keep sliding out of place. By cradling the head in your palm, or putting a finger in the kid's mouth (upper or lower jaw), you can hold the head in proper position while your other hand brings

the front legs and entire body forward with the doe's contractions. A kid puller (wire noose) can help if you have enough room to use safely. Be sure to sanitizer this puller before using, and clean it with mild soap and water afterward.

HEAD WITH NO FEET

Once you locate the front legs, you may need to push the head back in to aid in swinging the legs forward. Again, protect the uterine lining by cupping the hooves. I recommend you try hard to bring *both* legs forward to deliver the kid in the proper position. Using one leg and a head can work, but this is stressful on everyone, doe, kid, and human.

RUMP FIRST (BREECH)

Identify this abnormal position by feeling for the tail or following the rump structure to identify the attached hips and rear legs. Correct this position by pushing the rump back into the uterus. Slide your hand along one rear leg, feeling the hock and then down to the foot. Cup the foot with your fingers and bring it forward. Repeat on the other side to have both rear legs ready for delivery. Deliver the kid as a normal rear leg presentation.

FRONT FEET COMING, BUT EVERYTHING IS UPSIDE DOWN

Happily, this abnormal position is not seen often. You *have* to correct the position in utero. Do *not* attempt to deliver a kid that is upside down. First, determine that you have front feet upside down, *not* back ones right side up. Feel along the legs. If they are back feet, you will feel the angular hocks on the upper side of the legs, and the legs will flex down. Okay, you have front feet upside down. The trick to flipping the kid is well known to lifeguards. How do you flip a person who is face down in the water?

Push the fetus back into the uterus as much as you can. This maneuver needs space. If the uterus is jam-packed with other kids, you may want to get someone else out first. Twist your right wrist to the left as far as possible. Insert your hand into the uterus and grab the foot/leg that is on the left. In one sweeping motion, rotate your wrist to the right while moving forward with the kid. The kid should right

itself as you rotate your wrist. Once you flip the kid, position the head and feet correctly and deliver normally.

True story: I was working with beef cattle in Arkansas. We had a cow that needed help delivering her calf, which was upside down. The burly cattlemen couldn't get this calf turned. I had taken a life-saving class a few months prior and thought to try this turning technique on the calf. As I righted my wrist and pulled back, the calf immediately flipped and was out in no time! The call to the vet was cancelled, and I was taken to breakfast the next morning. Most importantly, the calf and cow were running the pasture together a few days later. I have also had success delivering a kid using the same technique.

SIDEWAYS

Occasionally, a kid will be presented ribs or spine first. Feel the direction of the ribs and determine which basic way the kid is coming, head or rump first. Reposition the kid to get either the front legs and head or rear legs and rump to come out.

All actions in helping the doe need to be done gently but with purpose. When moving feet, be careful to cup the hoof in your fingers to avoid tearing the uterine lining. When figuring out body parts inside the doe, look at another goat or close your eyes and mentally follow what you're feeling with your hand.

Zoonotic organisms (ones that can be transferred from animals to humans) can be in the birthing fluids. All goat keepers—particularly pregnant women and those with compromised immune systems—need to be careful when helping does deliver and during subsequent care of newborn kids. *This is very important if you are assisting a doe of unknown origin.* Wear protective gear, wash carefully before and after assisting, and take all precautions to minimize exposure to birthing fluids. Consult your health provider if necessary.

Birth Position	How to Assist	Warnings
Normal presentation: Front legs first, head next	Allow doe time to birth without assistance. If kid is getting hung up, add *slight* pressure by holding onto front legs and working with contractions.	*Be patient.* Assist only if kid or doe is showing signs of desperate distress. Clear nostrils and mouth of fluids when kid is born.
Normal presentation: Rear legs first, rump next	Allow doe time to birth without assistance. If kid is getting hung up, add *slight* pressure by holding onto rear legs and working with contractions.	*Be patient.* Be ready to assist if hips or shoulders get stuck. Clear nostrils and mouth of fluids when kid is born.
Abnormal presentation: One foot, head next	If doe is laboring without results, make sure foot and head belong to same kid. Follow head and find both attached feet. Correct as necessary.	Cup fingers/palm around hooves when correcting positions to avoid uterine tear.
Abnormal presentation: Two feet, no head	Follow both feet to make sure they belong to the same kid. Find the attached head, and correct position.	Difficult correction as head will keep slipping out of place. Hold crown of skull in palm *or* lower jaw with finger while moving kid into birth canal.
Abnormal presentation: Head, no feet	Follow head and find attached feet. Push kid back into uterine body to move legs into position.	Cup fingers/palm around hooves when correcting positions to avoid uterine tear.
Abnormal presentation: Breech – rump first no feet	Push kid into uterine body and gently reposition *both* rear legs.	Cup fingers/ palm around hooves when correcting positions to avoid uterine tear.
Abnormal presentation: Upside down	Push kid into uterine body. Twist right wrist to the left and grab leftmost foot. In one motion, rotate wrist to the right while pulling kid forward.	This maneuver needs some room in uterus. Deliver other kids first if necessary.

Opposite: Goats love mugging for the camera, especially at rest under the warm sun.

Chapter 8

Kid Care

Watching the goat kids grow from newborns to milking adults is just simply lots of *fun*! Yes, there are chores to do and problems that can arise, but young goats are a joy. They are silly and whimsical explorers, sprinting here and bouncing there. One minute they stand in serious thought, and in a flash, legs fly, bodies twist, heads charge forward—or sideways—or upward. Your work stops and you lose track of time. Oops. Kids have that mesmerizing ability to just stop the clock. These fluffy critters take away all cares and will have you laughing and dancing with them in no time.

Help them learn good habits and your style of life, and they will be a loyal and fun companions for years. What are good habits? Everyone will have a different opinion on this. There is a degree of discipline and routine learning that will create a safe, healthy, and positive environment for both animals and people. Cute goat kids grow up. Teach a kid to butt heads or push against you and in no time you will have a strong, large animal knocking you to the ground. These habits are hard to break once started, so don't start. Teach them to follow you inside their pen, or out for walks in woods. Many people bring goats into their homes, and some people claim to housebreak the goats to urinate and defecate in one spot. I am not in favor of this, since I believe that farm animals are best left outside, for their health and well-being. Goats are curious, and household items, such as electric cords or glass objects, can injure an animal. Some houseplants are poisonous. Set up a comfortable and safe pen for them in an outdoor building, and enjoy the animals in their environment.

Newborn kids are ready to have some fun! *Danielle Mulcahy*

Dairy animals are handled daily with milking chores. Teach young goat kids to come to the milking stand for attention, feeding, or treats. Also, begin at an early age to handle the feet and touch the udder area so hoof trimming and milking aren't terrifying tasks to the maturing goat.

Very young kids will demand your attention for feeding and care. If your family works full time, you may need to plan the birthing dates around vacation or holidays so you can devote several days to these "infants."

Newborn kids being bottle-fed thrive on feeding colostrum and milk three to eight times per day, depending on their weight and strength. Too much pampering creates a weak animal, in my opinion, but some newborns need the extra attention of frequent feedings, warmth, and TLC. A weak kid, once over the newborn hump and strengthened, can join in the group and grow as well as the rest.

I'm not a fan of heat lamps. An interesting warming jug, a heat barrel, is cheap and easy to construct. See construction advice at the Minnesota Meat Goat farm, www.vkvboers.com. This warming jug design uses a heat lamp. I've seen others that simply use a standard light bulb. The important point is to make it secure to remove the risk of fire. Another idea for warming cold kids is to make body warmers out of wool tights or sweaters. Place a hand-warmer packet in the wool "sleeve" to warm the kid. You can also use a microwavable flax or rice sleeve—if it is too hot for you, it's too hot for the kid. Very small and weak kids need immediate help. I bring them into a warm area (yes, sometimes the house) and make a towel-lined bed inside a Styrofoam box. A cooler or other type of insulated container will also work. Warm the towels in a dryer before wrapping the kid.

Very weak kids will need to be fed with a catheter tube. See the Tube Drenching Weak Kid Video at vimeo.com/user29715351. Here is the basic procedure:

How to Tube-drench Weak Kids

1. Retrieve a *clean* catheter tube and 60cc syringe from your kid box. Rinse it with warm water.
2. Warm the proper amount of colostrum or milk to be given at this feeding. The temperature should be lukewarm, feeling neutral when applied to your wrist. Weak kids should be given small amounts frequently. Often, their weakness also means slow digestive systems, and the milk simply ferments in their stomachs, causing a lethal gassy stomach/intestinal bloat. Depending on the size and weakness, I give 1 to 4 oz. of fluids every 2 to 4 hours. Receiving colostrum for the first 12 hours is very important. After this, I may switch to feeding electrolytes for a day to allow the stomach and intestinal system to strengthen, and then slowly start introducing milk.
3. Lay the catheter on the goat kid, simulating how it would travel down the esophagus to the stomach. Make a mental note of this length or have someone measure the length that will remain outside of the mouth.
4. Firmly attach the catheter tube to the syringe, and draw up the warmed fluid.
5. Remove any air in the syringe and tube by holding the syringe vertically with air trapped at the tip while keeping the catheter end in the fluid. Slowly depress the plunger to displace the air. Pull up additional fluid to fill syringe to the desired amount. Repeat until all air is removed. Keep the tube firmly attached to the syringe. **Don't put any unnecessary air into the kid's stomach.**
6. Place the kid in your lap and *hold the head level.* Carefully thread the catheter tube into the mouth and throat. Keeping the head level follows the natural flow of the esophagus. An elevated head opens the air passageways and runs the risk of putting fluid into the lungs. Thread the tube into the mouth until you reach the predetermined length.

Kid Feeding Schedule

	Colostrum	Milk or Replacer	Hay/ Forage	Grain	Salt/ Mineral	Water	Notes
Newborn to 2 days	4 to 8 oz. every 4 to 8 hours*		Available to nibble				Feed small amounts often
3–10 days		8 to 18 oz. every 8 to 12 hours	Available to nibble		Salt brick to lick		Feed to satisfy but not full
11–20 days		18 to 24 oz. every 8 to 12 hours	Free choice		Salt brick to lick	Available	
21–30 days		18 to 30 oz. every 8 to 12 hours	Free choice	Introduce 18 to 22% protein grain	Salt brick to lick	Available	Rumen begins to work
31–60 days		32 to 40 oz. every 12 hours	Free choice	18 to 22% protein grain, free choice	Salt brick to lick	Available	
60–70 days		32 oz. every 12 hours	Free choice	18 to 22% protein grain, ¼ to ½ pound, daily	Salt brick to lick	Available	
70+ days		32 oz. 1x/day for 1 week and *wean*	Free choice	18 to 22% protein grain, ¼ to ½ pound, daily	Salt brick to lick	Available	

7. Slowly depress the syringe plunger, discharging the fluid. The kid may startle with the sensation of fluids entering its stomach. Empty the syringe completely.
8. Slowly pull the tube out of the kid's mouth, keeping the tube attached to the syringe.
9. Repeat the procedure starting at #4 until the proper amount of fluid is given for the feeding.
10. Record how much was given at what time in your daily barn log.
11. Thoroughly clean the tube and syringe in warm soapy water and rinse well. Allow equipment to dry in a clean area.

Notes: *1 oz. (30cc) per 1 pound body weight every 4 to 8 hours, depending on size and strength. Do not overfeed. These amounts are for standard-size dairy goat kids weighing 7 to 10 pounds; proportion for smaller breeds and kids.

1. Keep kids head *slightly* elevated when feeding with bottles.
2. Wean at 75 to 90 days depending on weight and management.
3. Hay should be soft, leafy, and clean. Second cut is preferred. Introduce kid slowly to fresh forage. Be careful of poisonous plants, especially with spring growth.

Feed for Kids

The ingredients from the chart on page 106 are further explained in the following pages.

COLOSTRUM

Colostrum is the first milk produced by the doe when she gives birth. Colostrum is thick, creamy yellow in color and full of passive immunity goodness. The doe's protective immunoglobulins cannot pass to the growing fetuses through the placenta, so this rich, antibody-laden colostrum is very important for newborn goats. Just prior to parturition, the doe's body increases production of immunoglobulins (antibodies). These protective proteins are secreted in the colostrum. With the initial suckling of colostrum, on doe or bottle, the doe's immunoglobulins are absorbed through the

Label colostrum, chill, and freeze for future use. *Jay Iversen*

kid's intestinal system into the blood. Older does generally have colostrum with a wider range of immune protection, as they have been exposed to more organisms. The doe's initial colostrum contains the highest concentration of immunoglobulins, corresponding with the newborn kid's ability to absorb such nutrients. These facts highlight the importance of making sure the first colostrum is drunk within 20 minutes to 4 hours after birth. Two to 8 oz. of colostrum is recommended, depending on birth weight. If you are allowing the kid to suckle on the doe, I recommend milking some colostrum out of the doe and bottle feeding it to the newborn kid as soon as possible to ensure adequate consumption of colostrum. You can use this chance to also 1) make sure the doe has good milk flow, 2) help teach the kid how to nurse, and 3) introduce bottle feeding to a newborn kid. (See the CAE Prevention sidebar on page 95.)

The doe will continue to produce colostrum for the next couple of days with complete transition to useable milk by day four for most does. Research shows that goat kids can absorb immunoglobulins for 3 to 4 days after birth.[2] Ideally, feed kids with *their* mother's milk for the first four days so they receive the best protection of colostrum passive immunity.

2 *Goat Medicine*, p. 308

Colostrum, with its high percentage of proteins, needs to be warmed slowly. Do not use a microwave or boiling water. Temperatures above 140° F (60° C) will denature the antibodies, destroying the immune protection. Higher temperature will also turn the colostrum into a thick and undrinkable "pudding."

Dairy goats generally produce more colostrum than their newborns can consume. Gather colostrum from a doe having extra, and store it in containers labeled with the doe's identification, the date, and the number of days after birthing in a refrigerator at 35 to 40°F (2-5°C).

Thoroughly chill colostrum, and then freeze it for longer storage. Colostrum can be stored in plastic freezer bags or containers (bottles). I make colostrum cubes. Simply pour chilled colostrum into ice cube trays and freeze. Pop the frozen cubes and store in a labeled freezer bag. Cubes warm quickly and in small amounts so you don't waste stored colostrum.

Some people use colostrum for human consumption and skin ointments. Please make sure your newborn goats have adequate amounts and you have a frozen stockpile before taking colostrum for your own use.

Feed goat kids milk for 10–12 weeks after birth.

MILK OR REPLACER

Goat milk is the best milk to feed your growing kids. If you don't have goat milk, milk replacer or cow milk are options to consider. Regardless of what you choose to feed, *always* transition the goat kids gradually to a new kind of milk. When purchasing young kids, ask the breeder for milk from his or her farm when you negotiate the price. Some farms include milk in the purchase price of the kids, while others will charge you. Bring your own *clean* container for the milk.

The first feeding of purchased kids should be 100-percent milk from the birth farm. Over the next 3 to 4 days, adjust kids to your chosen milk. Start the transition with ¾ birth farm milk plus ¼ of the milk you've chosen. Move to ½ and ½, ¼ birth plus ¾ chosen, and then to 100-percent chosen milk. If you opt for milk replacer, I suggest using a high-quality goat milk replacer. Do not use lamb milk replacer, as this has too much fat and often causes diarrhea in goat kids. Calf milk replacer can be used, but be observant for steady growth and any signs of diarrhea. If problems develop, switch to goat milk or goat milk replacer, and slowly transition when the kids are strong and doing well.

HAY/FORAGE

A soft, second- or third-cut *mixed* grass, clover, or alfalfa hay is preferred for young goats. I avoid pure alfalfa hay as this can be too rich for newborns and can cause diarrhea. Offer the hay as free choice for the first 2 to 3 months. Watch how much the goats are eating and adjust the amount as they age. The amount fed varies greatly depending on other feeds available. I encourage young kids to go outside and forage as soon as safely possible. Issues of safety include weather conditions, predator control, forage growth to minimize parasite infestation, growth of poisonous plants, and fence security. Kids mimic their moms and will be nibbling at hay within a day or two if you are raising kids with the does. You may have to teach separated kids by pretending to eat the hay yourself. The kid's rumen begins to function around 3 weeks of age. Having hay and forage available for nibbling as soon as possible encourages rumen development and function. Place hay up and out of the pen to keep free of feces, urine, and water. Do not feed moldy hay.

Feed a soft second- or third-cut mixed hay. *Shutterstock*

GRAIN

Ask five people and you will get five different answers on what if any grain to feed to kids. Read resources, ask goat mentors, and attend educational sessions. If you are feeding grain, I recommend starting with an 18- to 22-percent protein grain. Introduce free choice in small amounts around 3 weeks old. Use adjustable-height mangers or tubs that can be easily changed. The goat kids should be growing at a steady rate—what works this week will be too short in 10 days. Best if these can be outside of the pen so the kids do not step, urinate, or defecate in them. You also don't want to waste grain, so give in small amounts. By 2 months old, they will be eating hay and grain steadily. Depending on their weight and growth, feed kids ¼ to ½ pound of grain split into two or three feedings. Medicated feed is not necessary with proper care and cleaning of stalls. See the Kid Health section on page 115 for information on causes of diarrhea and how to prevent and treat it.

SALT/MINERALS

Goats in general have a higher need for minerals and salt than other farm animals. I find young kids start licking salt and free choice minerals as early as a week old. I like using the small salt bricks or hanging donuts for kids. Licking the brick keeps the kids occupied for a few minutes each day.

WATER

Young kids are *very* active! They need water. A water bucket is best placed outside the pen so they don't soil and spill the water. This may be easier said than done. Regardless, change the water at least twice a day, refreshing with clean, warm water.

Depending on your goals and time, keeping track of growth rate is important. Weigh the kids when they are born by either using a sling scale or simply holding them in your arms while standing on a scale. Record the date and weight. Do this once a week to track kids' growth progress. Some medications are dosed by weight. **Do not guess**—even if you are experienced. Weight of goat kids varies depending on many factors, including bone structure. Being off by 5 pounds is a 25-percent error for a 20-pound kid. This is huge for weight-dosed medications.

A big challenge for beginner and experienced goat raisers alike is setting up a pen for goat kids. The kids grow so fast that keeping up with changes for the first couple of months can be tricky.

Kids Raised with the Does

Provide a creep area where the kids can hide and sleep safely away from the does. Place a simple, sturdy wooden platform 2 feet in height against a wall and you have both a "cave" underneath where the kids can sleep and a shelf for both adults and kids to stand and play. (See illustration on page 16.) Use sense to determine how many of these areas to have in your pen, depending on your total herd number. Be sure to move the platforms and clean these areas every time you clean the pen. Place dry bedding in the "cave" area for warmth and health of the kids. Watch for loose wires, sharp objects, and small areas where kids can be trapped or injured. Also, check for areas where predators or stray dogs can nab kids. I've seen kids die with injury from dogs chewing on limbs that were dangling from raised pen structures. Some things are still learned from experience, and something like that need only occur once to demonstrate the error in a pen structure.

Kids Raised Separate from Does

Kids like to sleep in a protected area. The same platform/cave system used in a pen with mature animals is loved by kids raised separate from their mothers. The platform gives kids a place to jump to and from. Placing a few in a pen creates a fun play area. Lengthen the distance between the platforms as the kids grow to give them a challenge. We use the large plastic dog carriers in pens, leaving the door open (or removing it completely) so the kids can go in and out as they please. Again, clean these sleep areas frequently and bed with dry shavings, sawdust, or a good depth of straw or hay.

The most important part of having kid pens are to keep them clean, safe, draft-free but ventilated, well lit, and allowing the kids room to move. If your pen is smaller, please take time to let the kids run and jump several times every day. These little critters are so much healthier and grow better if they get the time and space to play.

Use your imagination and sense when setting up the area for feeding milk, hay, water, and grain. A simple multiple-bottle holder can be made cheaply and with little skill.

A board with a single nipple attached to a long tube sitting in a jug of milk can feed a pen of many kids all day long. Fill it with cold milk so the kids will not drink too much at one time. Change and scrub the nipple, tube, and milk container daily. Feeding a group of kids with a bucket feeder or "lambar" is easy. We suspend the lambar on a chain, adjusting the height easily as the kids grow. Be sure to rehang the chain out of jumping kids' height when not using it. Lambars can also be set in a holder, placed on blocks for proper height of suckling kids.

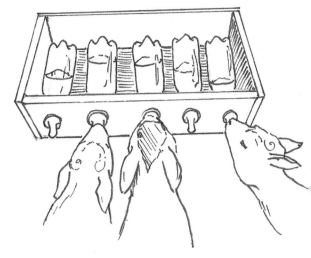

A simple multiple bottle holder. *Danielle Mulcahy*

Lambar or bucket feeder. Hang from a chain or set in a holder.

Some kids will drink directly out of a pan or bucket. Older references say pan feeding does not allow the esophageal/reticular groove to properly form and suggest bottle feeding only. I have done both with good results, but I prefer bottle or lambar feeding for goat kids. Regardless, measure the amount being fed, and don't let kids drink too much, as this can cause diarrhea and other health issues.

Training kids to the electric net fence is most successful when kids are trained early. I begin training around 2 months of age. Set an electric net fence in an area that also has a physical and visual barrier, such as a wooden fence or walled area. Goat kids tend to jump forward. These barriers help train the kid to step back instead and also contain the kids if they do push through. Test the voltage of the electrified fence to make sure it is properly charged. Most fences require at least 3,000 volts. Have several people on hand for the initial lesson, including one to quickly unplug the electric charger if a kid gets tangled in the fence. The curious kids will approach the fence. One will investigate closely and get snapped. Another may also get zapped. A third may possibly try, but after this, the kids

realize this is something they need to respect. The kids have learned to not challenge the fence. Certainly keep a close eye for the next several days, and be aware that they are young animals and need supportive care. Keep the net fence in good condition and free of plant growth. A second training may be needed when you move the kids to a new location with a new fence setup.

There are several health issues with young goat kids. Prevent illness by giving kids a good start with a clean birthing area, colostrum as soon as possible, proper feeding, a frequently cleaned kid pen that has good ventilation without cold drafts, plenty of exercise in a safe area, immediate attention to health concerns, and lots of TLC!

Following are some brief notes on kid health. See Chapter 9, Health Care, for detailed information on assessing goats and Chapter 4, Management, for additional kid information. Remember to write information in your barn log, specific animal details in individual goat records, and scheduled tasks on your barn calendar. Goat kids grow up so fast. Enjoy their antics and playfulness. Nothing takes the stress of a day away like laughing at goat kids.

Train kids early to respect an electric fence.

Top and bottom: Once they know their boundaries, goats will be quick to play and quick to rest.

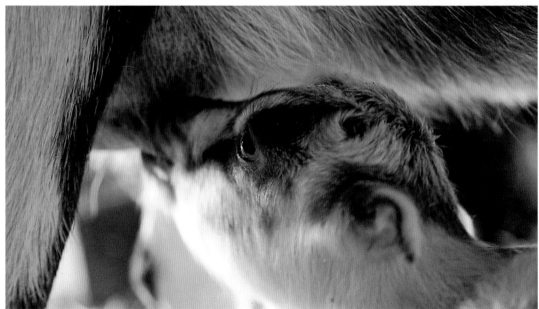

Top: Newborn kids like to sleep together for both warmth and comfort. **Above:** Mother's milk is the best treat for healthy kids.

Danielle Mulcahy

Kid Health
VACCINATING

The primary vaccine used for goat kids is Clostridium C/D and Tetanus (CD/T). The C/D stands for strains of *Clostridium perfringes* C and D. These are the common strains that cause enterotoxemia (severe intestinal problems) in goats. Symptoms are diarrhea, lethargy, kicking at the stomach and crying, and sometimes bloat. The organisms can grow at various speeds, from extremely fast and deadly to slow and chronic. The clostridial organisms produce a toxin that damages intestinal tissue. Tetanus is also caused by a clostridial organism, *Clostridium tetani*. Newborns receive a passive immunity to these organisms through drinking colostrum, if the does have been previously vaccinated. By 1 month of age, the kid's immune system is developed enough to begin making its own antibodies. Vaccinate kids with 2 cc C/D and T vaccine (sub-cutaneous) when they are 1 month old. Repeat with a booster shot in 3 to 4 weeks. Record the date of vaccination on kid charts. This vaccine has a 21-day meat withdrawal.

Antitoxin is available for both enterotoxemia and tetanus. This provides short-term protection and works quickly. The antitoxin is recommended just prior to castration and tattooing. Antitoxin is given primarily to young goats that have not yet built up a working immune system.

Vaccinating your goats will help keep them healthy for a long time.

Diarrhea and respiratory problems in kids is common, but easily treated. *Shutterstock*

Diarrhea in young kids may be caused by coccidia, a single-celled protozoan that is naturally present in goats' gastrointestinal systems. As kids grow and develop their immune systems, the natural flora is kept in check. Generally, at 3 to 4 weeks of age and again at 7 to 8 weeks of age, the coccidia growth can outpace the kid's ability to fight back, causing diarrhea, lethargy, and general ill health. This outbreak can often be traced to unclean pens or soiled hay, water, and grain feeders. Keep kid pens clean to minimize this problem. Fortunately, there are effective treatments available. The two most popular medications are amprolium and sulfadimethoxine. Use as recommended on the package. Kids should show improvement by the first or second day of treatment. Continue treating as directed. Keep in mind the meat withdrawal period for these medications. Often, if one kid in a group breaks with diarrhea, the others will follow. Group treatment with these medications may be advised, based on circumstances. Consult your veterinary professional or goat mentor to help diagnosis and come up with a proper treatment. Coccidia can cause permanent scarring of the small intestine, resulting in poor growth and production. Prevent it with proper husbandry of feeding colostrum, regular pen cleaning, and plenty of fresh air and exercise for growing kids.

Other causative agents for diarrhea in young kids include salmonella, *E. coli*, rotavirus, and *Clostridium ssp.* See the resources on page 180 for more detailed health information.

Respiratory problems can affect young kids as well. Be sure their pens have plenty of airflow without cold drafts. Damp conditions at spring kidding can generate pneumonia. Weak kids may have difficulties recovering, especially because the causative agents are hard to identify. Often, treatments are selected based on herd history and experience, with the real cause never being known.

A doe's health is a good indicator of her kids' health—sick does can produce tainted milk.

Chapter 9

Health Care

Goats are generally one of the healthiest and most adaptable of domesticated animals. Proper care from the time they are born sets the stage for a well-balanced goat. The care of dairy goats includes access to nutritional feed and forage, adequate shelter, minimal stress, and hygiene at milking. Preventing illness by promoting health is the best approach. Previous chapters helped teach you about raising healthy, strong goats. Set up a good housing area, create a sound chore routine with caring and knowledgeable people, provide nutritious feed, keep useful records, and use common sense. I encourage you to read and learn about dairy goat well-being, take notes, and keep a barn log with your own experiences, good and bad. As previously stated, spend time with a goat mentor. Most importantly, watch your animals. Take your clues from them. Animals in general and dairy goats in particular can teach you a lot if you have the patience and understanding to listen. Be curious.

Observe. There's an old farmer adage: "There is no such thing as a sick goat; they're either dead or alive." This is true, to a degree. Often goats do not display sickness until they are past the point of no return. This is a survival skill. Noticing any changes in your goats' behavior and acting on those changes is important. The eyes hold great clues to a goat's wellness. Dairy goats have the advantage of needing to be milked daily, so you will get to know your animals and their eyes quite well in a short time. Use this to your advantage. Plus, remember to keep watch on bucks and kids. They can sometimes be forgotten in the back shed. Get in the habit of quickly assessing your animals every day.

Act. The more you practice assessing your goats, the more adept you will become. A purposeful glance will tell you plenty. Get a good understanding of normal. If something looks amiss, spend a few minutes looking closer. If it's only more puzzling, grab some vital tools, such as a thermometer and stethoscope, and learn more. Write notes, take pictures, and create your own health library.

My approach to animal healing is mixed with traditional veterinary care and complementary treatments, such as herbal teas and tinctures and homeopathic remedies. I assess the individual animal, taking into account the immediate concern, the animal's age, history, and specific personality, and in a business sense, the dollar value. We'll talk more about this last subject later.

In this chapter, you will learn how to assess your dairy goat. Following will be some specific health concerns, the Top Ten on page 126. Next, I'll present brief information about common herbal remedies and homeopathic veterinary care, and then conclude with some thoughts on the economics of illness and death. Goat healing is covered in many good books and Internet sites. See the reference section on page 180 for some of my favorites.

Remember that goats are living creatures, and despite all your goodness, complications do occur. Life is a cycle of birth and death. Be realistic and understanding in your expectations of recovery from illness. Learn from experiences and mistakes.

Second, a good relationship with a licensed and practicing veterinarian is important. Happily, there are more veterinarians today who are interested and knowledgeable in goat medicine. Find a local vet to visit your goats annually. Ask meaningful questions. Make a connection prior to needing assistance during a crisis.

Third, you are working with dairy animals that produce milk as a food product. Many commercial medications require a milk withdrawal period, but few are labeled for goats. Work with your veterinarian or an experienced goat mentor for advice on milk withdrawal when giving any medications, including deworming anthelmintics.

Dairy Goat Assessment

(For more information, you can view the video for Health Assessment at vimeo.com/user29715351.)

STEP 1: OBSERVE APPEARANCE

Observe the goat's behavior. Is she acting normal? Observe the hair coat, body structure, eyes, ears, limbs, and udder. Is the hair coat shiny and "strong"? Are the eyes bright and alert? Is there any nasal discharge or coughing? Is there discomfort or hesitation to bear weight on any limbs? Is the goat able to stand and walk? Are there unusual vaginal discharges? What about diarrhea or problems urinating? Is the udder hard, swollen, or tender? By observing your goats daily, you can notice any changes, even subtle ones.

If you observe an obvious cut or limb injury, apply appropriate first aid to stop the bleeding and cleanse the wound (keep reading for physical injury first aid). Otherwise, if something is amiss, move on to the next step.

STEP 2: TAKE VITALS

Use the following four vital body functions to learn more about your goat's health.

Temperature

The goat's temperature is an important indicator of the animal's health. Take the temperature *any* time you think something is wrong. Normal temperature is 101.5 to 104°F (38.6 to 40°C). There are two main types of thermometers: glass and digital.

Have both in your health kit. I prefer the glass thermometers. While they take more time and have the potential to break, I think they are more accurate and reliable. Attach a string and clip to the glass thermometer to reduce risk of loss or breakage. Shake down the liquid in the thermometer so the indicator is below 98°F (36.7°C). Apply lubricant to the thermometer, insert the instrument into the animal's rectum, and attach the clip to its tail hair. Continue to assess the goat while the temperature settles—a minute plus will do the trick. Digital thermometers are an easier and faster way to take a temperature. To use a digital thermometer, simply apply lubricant, click the "on" button, and insert the instrument into the animal's rectum, and wait for the thermometer to beep with final reading. A higher than normal temperature indicates some type of infection or inflammation in the goat's system. A hot day in the sun, a chase to be caught, or other activities can raise the goat's body temperature as well. A lower than normal temperature can mean the goat's system is shutting down. Immediate supportive care, hydration, and possibly moving the animal to a comfortable, heated area may be necessary.

Rectal thermometers are simple tools to help assess the health of your goat.

A cold day or a lot of manure being passed can also lower the temperature. Consider the goat's recent activity to help determine the cause of its abnormal temperature and how to proceed.

Respiration Rate

You can easily tell if your goat is breathing fast or slow by simply watching and gently listening. Look at and listen to your goat now and again when it is healthy to learn the normal pattern so you can tell when something is wrong. A stethoscope is best to determine the actual breathing rate. Normal is 12 to 15 breaths per minute, with young goats having a faster rate. Determining respiratory issues, such as pneumonia, in the lungs, and pinpointing specific areas of congestion takes concentrated instruction and practice. Changes in respiratory rate can be due to lung infection, metabolic imbalances, injury, dehydration, parasites, stress, or ingestion of poisonous material.

Heart Rate

This rate is also best determined with a stethoscope. You can place your hand on the bottom of the goat's chest and feel the heartbeat. A normal rate is 70 to 80 beats per minute (bpm). As with respiratory rate, this rate is faster with young goats. Activity will obviously increase the heart rate. Specific heart issues require instruction and practice to hear and understand. Changes in heart rate can be due to metabolic imbalances, dehydration, stress, pain, or injury.

Rumination Rate

A healthy rumen is quite active with churning and mixing the contents in a regular, rhythmic cycle. You can hear this activity by pressing your ear against the goat's left side. The normal rate is one to two churns per minute. A stethoscope makes this activity quite entertaining, especially for children and visitors. Goats with metabolic imbalances, gastric upset, difficulty with birthing, and general malaise can suffer with rumen disturbance.

Record these vitals in your barn or individual animal log, even if they are normal. The fact you moved to taking vital sign readings indicates something wasn't quite right. Brief notes may help if, in a few days, something else is wrong with that animal. Include date and time and even weather if you think it may be a reason for changes in your goat's behavior.

Continue examining the goat, taking into account the animal's or herd's recent history, such as breeding or kidding, change of feed or browse area, escape from fenced paddock, or recent return from a show or travel. Also consider the animal's age and individual health history.

STEP 3: INVESTIGATE INTAKE AND OUTPUT

The next step is to observe swallowing, urination, and defecation. The body's intake and discharge systems have to work. Problems with swallowing can be due to irritation of the lips, tongue, gums, and teeth. Goats are notorious for getting pieces of brambles or sticks lodged in their mouths. Swollen areas and abscesses of pus can develop, causing discomfort and blockage. Use a 6- to 8-inch long piece of 1-inch PVC to gently wedge the mouth open to look for problems. Be careful; the teeth are very sharp, and a goat doesn't like you prodding around in its mouth. Salivary gland cysts are also prevalent in goats. Generally they will rupture externally and little fuss is needed. Goats, particularly as they age, may develop problems with broken and missing teeth. Broken teeth can be due to excessive gnawing on hard objects such as posts and gates. This chewing can simply be a habit, show boredom, or, more seriously, indicate a mineral deficiency. Try to correct it by putting out minerals, gradually changing your feedstuffs, introducing an animal buddy, taking the goats for a walk, or changing the play objects in their exercise area. Goats like challenges—give them something to do!

The most common problem with urination is in males, especially castrated ones. Primarily, crystals of calcium phosphate, calcium apatite, and magnesium ammonium phosphate (struvite) form in the urinary tract—sometimes forming stones, or urinary calculi—and block the outflow of urine in the urethra. This is very painful and left untreated will rupture the urethra or bladder, leading to death.

Vitals and Other Important Information

Vital Sign	Normal Readings
Temperature	101.5 to 104°F, 38.6 to 40.0°C
Heart Rate	70 to 80 beats per minute, faster for kids
Respiration Rate	12 to 15 breaths per minute, faster for kids
Rumination Cycle	1 to 2 churns per minute
Water Consumption	1 to 6 gallons/day for a mature, lactating goat, depending on many factors (see Chapter 3, Feeding)
Estrous (Heat) Cycle	Average 21 days, range of 18 to 22 days
Length of Estrus (Heat)	2 to 72 hours
Gestation (Pregnancy)	Average 150 days, range of 145 to 156 days
Longevity	10 to 12 years for goats in a commercial herd, 12 to 15 years for goats in a smaller herd

Changes in the frequency or consistency of fecal matter are important to assess. If necessary, you must isolate a goat to carefully observe its individual fecal output. Diarrhea or loose stools may be evident by soiling of their rectal area, tail, and rear legs. These changes can be due to many factors, so determining if treatment is necessary takes observation and experience. Has the feed changed? Are there recent changes in hay, the pasture/browse area, grain, or supplements? Did a neighbor throw shrub clippings over the fence? Some common yard plants, such as yews and rhododendrons, are poisonous. How terrible for a well-meaning individual to accidentally poison an animal!

Instruct *anyone* who may have access to your animals not to feed your goats unless you approve the plant material!

Are parasites an issue? Did you properly quarantine that new animal and check for resistant parasites? Do the young kids have a coccidia overload? Is there a recent stressor? Issues of feed changes and stress generally just need some time and gentle grass hay to correct. Parasite and acute issues such as poisoning need attention, often immediately.

Stool samples may need to be submitted to a diagnostic laboratory to identify the offending organism and start proper treatment. This takes times, so supportive care and treating the potential cause may be necessary while waiting for results. Dehydration caused by diarrhea is a critical issue, especially for a dairy animal. Administering electrolytes or beneficial herbal teas is a good step for supporting the goat while the intestinal tract is healing. Permanent damage/scarring to the intestinal lining is a possible risk with prolonged diarrhea, as is death. Don't overlook this symptom and just assume it will resolve on its own. Assess your goat further and observe it carefully for signs of weakness and dehydration.

 Remember that you determine left and right sides of an animal by standing behind it and looking at the back of its head.

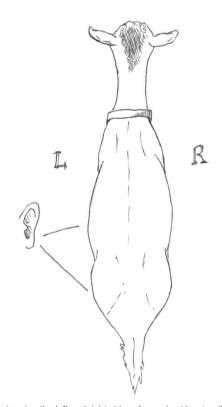

Determine the left and right sides of an animal by standing behind it and looking toward the back of its head.

STEP 4: CHECK FOR BLOAT OR RUMEN DISTENTION

The rumen is located on the goat's left side. Stand the goat up and observe the left side behind the ribs. Imagine a big balloon inside the rumen. This is what bloat looks and sounds like; the goat's left side is distended upward and out. Ping the skin with your fingers—the sound is tympanic, and the feel is taut. Bloat can be caused by several agents but is primarily feed related. The cause can be difficult to pinpoint. If several animals experience bloating, look hard for the cause and make changes. Occurrences here and there can simply be caused by a particular animal's weakness. Bloat is a gas buildup in the rumen and can be quite dangerous, even fatal. Act *immediately* to relieve it.

STEP 5: EXAMINE EYES AND EARS

Eyes are the window to your soul. Yes. Look into a goat's eyes and you are connected to something special and unique. Look closely to also determine its health. The eyes themselves should be bright and clear. Twitching eyes indicate a nervous system difficulty or perhaps a poisoning. The causative agent may be hard to pinpoint. Poor vitamin B absorption, rabies, meningeal worm, and metabolic disturbances are all possible causes of eye issues. A thiamine (B1) or B complex injection can resolve the issue or at least support the system while you identify the causative agent. Work closely with your veterinarian as some causes, such as rabies, must be reported and are of concern to human health. The muscle contraction of the pupil is a good indication if there is a problem with a mineral imbalance, such as systemic levels of calcium, phosphorus, and magnesium.

Carefully shine a flashlight at the goat's eye if you suspect hypocalcemia (milk fever). The pupil constriction will be delayed or slow if this mineral imbalance is present. An eye injury, such as being poked with a stick or stray piece of wire, can appear concerning as it heals since the cornea turns cloudy blue to white as it begins to heal from a scratch or other minor injury. Blood vessels grow on the outer surface to continue the healing. Many mistake this cloudy mass as the injury itself, while it is actually the body shrouding the eye to promote a quick return to normal. However, the injured surface can become infected before proper healing can take place. Eye ointments, such as tetracycline, can quicken the healing process. If there is pus present or the goat spikes a fever and acts abnormal, seek professional help immediately.

The membranes under the eyelids (conjunctiva) have active blood flow. Pale membranes may indicate a loss of red blood cells (anemia). This is the basis for FAMACHA, a technique of assessing parasite levels.

A drooping ear or ears can signal a nervous system problem, specific causative agent, difficulty or blockage in ear/sinus/lymph system, or a metabolic disturbance. A pathogenic bacteria, *Listeria ssp.*, can damage the brain and nervous system, causing the appearance of paralysis on one side and a circling gait. A meningeal worm infection can also cause drooping ears.

STEP 6: EXAMINE GUMS AND NOSE
Gums
Along with the paleness of eye membranes, pallid gums indicate an anemia problem. Some animals simply have pale membranes without experiencing a problem. Get to know what's normal and record it in your individual animal records. Changes in normal are indicative of a problem. Practice by observing the color of the gum, and then gently press and hold a finger on this gum area for 3 to 5 seconds. Release the pressure and see how quickly color returns to the gum. Heavy paleness or a slow return of color indicates a blood anemia issue.

Another consideration of the gum condition is dehydration. The gums should quickly respond to normal tension when depressed for color observation. Dry, slack gums are indicative of dehydration. A compromised goat has a slow response to healing if not properly hydrated. This is especially true for metabolic issues. A healthy, mature dairy goat that is lactating needs 1 to 6 gallons of water daily.

Nose
Your goat's nose should be moist and soft, without drainage. Dry, scaly conditions can indicate dehydration or mineral needs.

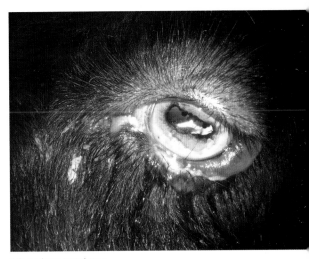
Infected eye membrane.

Excessive drainage may be from infected sinuses, respiratory issues, or simply a minor congestion. Consider the animal's history, weather conditions, and previous assessment steps, such as the respiration rate or lung sounds. As stated previously, goats are susceptible to lung infections. A compromised respiratory system can stress the entire body, leading to more complications.

STEP 7: FEEL THE UDDER
The udder, or mammary gland, should be supple and smooth. If the goat is lactating, milk should flow easily from a teat with proper milking technique.

The udder should be warm as with normal body temperature. Indications of mastitis (mammary infection) include hard and knotty tissue and milk that is clumpy, stringy, bloody, and possibly foul smelling. The udder may feel hot or even cold. The goat may have a high temperature, not want to eat or come in to be milked, and have general malaise. Mastitis comes in varying degrees of severity and success with healing.

Other concerns with the udder include cuts, abrasions, and teat trauma. Happily, the udder is quite vascular so healing is generally quick. Apply standard first aid with frequent cleansing of the wound and proper dressing. Teat trauma is difficult to heal, as the skin is disrupted twice daily with milking, and is sensitive (AKA the goat kicks). I have used herbal ointments with calendula and comfrey to promote tissue healing. Also, the milk may be slightly bloody if the goat has been fighting or had other trauma. Goats with this condition do not generally have other symptoms of mastitis. Discard milk for a few days until the affected tissue naturally heals and milk is normal. The udder's halves are anatomically separated, meaning the milk from one half does not flow into the other half. If milk from one half is abnormal, milk from the other half can be used **unless you treat the animal with any type of medication**. Medications travel through the blood stream and will be present in milk from both udder halves.

STEP 8: CHECK VAGINAL DISCHARGES

An active, functioning reproductive system is needed for a dairy goat to successfully get pregnant, give birth, and begin producing milk.

Healthy vaginal discharge is a clear, thin mucous with no odor and is generally only seen during the active breeding season. Thick, creamy white discharge with a pungent smell is indicative of uterine infection. Treat uterine infections before breeding a doe or taking her to a buck for breeding. Some infectious organisms can be spread through breeding. Bloody vaginal discharges following routine births are normal. The uterus begins to shrink and return to its smaller size quickly after the kids are born. Pregnancy membranes, blood, and fluids are expelled during this process. There may be an off smell to these discharges, but watch the doe for a high temperature and malaise before treating, as many of these situations resolve without intervention. Retained placentas infections due to birthing difficulties, and other maladies do need to be addressed.

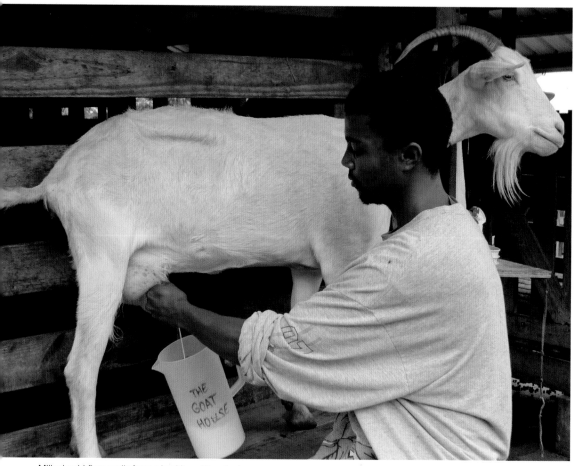

Milk should flow easily from a healthy udder. *Jay Iversen*

STEP 9: FEEL FEET, LEGS, AND SKELETAL SYSTEM FROM HEAD TO TAIL

Goats are capricious! They are quite agile and generally leap and bound with endless energy and accuracy. Changes in your goats' normal activity level may be due to a musculoskeletal injury, bone misalignment, or joint affliction. Get in the habit of running your hands over their bodies—head to tail, spine to hoof—to feel their normal body frames. The more you know about goat anatomy, the easier (and cheaper) it will be to maintain your herd.

As with other aspects of goat health, prevention is the key. Inspect their pen and yard regularly for loose and sharp metal, nails, boards, holes in the fence, and other concerns that can cause injury. While it's amazing what duct tape and baling twine can *temporarily* fix, be aware that goats will quickly undo tape and twine, and possibly try to eat it—causing more problems!

First Aid

Important: *Consult a veterinary professional when your experience and abilities are not sufficient to handle an illness. Keep accurate records, ask questions, and listen. Many veterinarians are happy to teach. Take time and learn.*

If a goat is limping, carefully follow the bones in the affected limb to determine if there is a break or swelling. Check all limbs, as goats can fool you with injuries. If there is no obvious injury, you will need to look further. Limping on forelimbs can be caused by soreness due to injuries in the rib and chest area. The hip and pelvic area can be misaligned during a difficult birthing. Check hooves for punctures wounds, cuts, or breaks in the hoof walls. Goats can be bullies, and injuries do occur. Once again, injuries will be minimized if you keep your goats well fed with balanced nutrition, make sure they have exercise and play time, and create a safe shelter and environment for them.

Broken bones will need to be reset and carefully splinted. Many veterinarians have

Knowing your goat from head to tail is tantamount to an enjoyable—and hopefully profitable—dairy goat experience. *Shutterstock*

portable x-ray machines. I keep packing gauze and splints of varying sizes on hand. Be careful to set bones correctly, give adequate support with a properly sized splint, and apply wrapping with the right amount of tightness. Check the area frequently for swelling, infection, heat, blood flow, or irritation of the splint. Leave the bone splinted for 6 to 8 weeks, redressing as needed. A compound fracture (broken bone sticking out of skin) will need extra care to be properly set and treated for infection. Seek professional help to minimize risk. Slowly reintroduce the animal to the herd so the healing goat doesn't have further injury by other goats pushing it around.

Cuts can be cleansed with an iodine solution wash diluted to the color of weak tea. Rinse with clean water, and dry thoroughly before bandaging. Cover the wound with gauze or first aid bandage. Wrap the area with an elastic bandage, and secure the bandage with electrical tape.

Depending on the injury, cleanse the wound and apply a fresh bandage every 12 to 24 hours for the first 2 or 3 days, and then every 3 to 5 days. Keep a close eye on the wound and the animal's general health, checking for swelling, discharge, or heat near the injury. Address any indications of inflammation immediately by cleansing and redressing the wound. Punctures and deep wounds need to be addressed daily, and it's vital that they heal from the inside out, or they may form an abscess. A sterile wick may be needed to induce healing in such cases.

Goats are sensitive to tetanus. These types of injuries underline the importance of vaccinating annually with Clostridial CD/T vaccine.

Trim and fix injured hooves on a case-by-case basis. See the information on basic hoof trimming on page 66 of Chapter 4, Management. Some injuries to tendons and ligaments cannot be healed.

Goats are survivors—it is amazing how quickly and thoroughly they heal with proper first aid and supportive care.

TOP TEN COMMON GOAT ILLNESSES
1. Parasites

Parasites are currently a topic of much concern, with advances in diagnosis, prevention, treatment, and general assessment of both external and internal parasites. In general, parasites are organisms that feed and grow on or in a different organism or host. Parasites do not provide any benefit to the host and often cause harm or weakness.

External parasites—most commonly lice, fleas, and mites—are generally not a major problem for goats. These organisms are generally associated with overcrowding of pens, soiled bedding and ill-kept pens, and poor nutrition. Getting out of doors and into the sunshine will help rid goats of lice, which are often seen on goats after a long, cold winter. Louse powders are available, but use them carefully with lactating animals, and check for a milk withdrawal. Herbal rubs of eucalyptus oil can also help. A spray of hydrogen peroxide

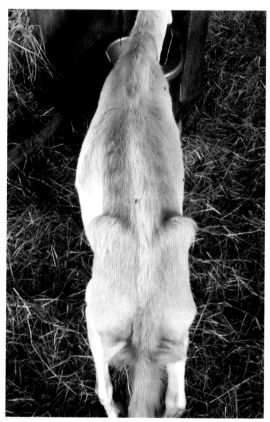

Doe in poor body condition, possibly due to high parasite levels.

will help alleviate mites, which cause flaky, dry patches on the skin, often in the ears, tail head, and feet. Consult with your goat mentor or veterinarian if you have stubborn external parasite problems.

Internal parasites are more complex and troubling for goats. Through the years, internal parasites have developed resistance to commonly used dewormers (anthelmintics), resulting in strains of parasites that are difficult to control. Common internal parasites include barberpole (large stomach) worm (*Haemonchus contortus*), common threadworm (*Strongyloides papillosus*), small intestinal worm (*Coopera*

spp.), and goat tapeworm (*Moniezia expansa*). These parasites ingest blood from the host to thrive and reproduce. See detailed information on individual parasite life cycles in references listed in the resources on page 180.

Prevention, along with keeping goats strong and nutritionally sound, is the key to controlling parasites. Many areas in the United States and abroad have weather conditions that favor parasite growth. Hot climates, irrigated pastures, limited browsing lands, and historic overcrowding conditions can generate parasite problems. Follow these considerations to minimize parasite issues.

Minimize parasite larvae ingestion by feeding in area with plant growth at least 4 inches in height.
Danielle Mulcahy

Goats love willows.

Parasite Test Recommendation

Seek advice on the best procedures for parasite control in your area. The recommendations in our area of Massachusetts call for two negative fecal tests (less than 500 eggs per gram, EPG) taken 2 weeks apart. If treatment is indicated in quarantine animals, ensure a 90 to 95 percent reduction in EPG following treatment.

This indicates no resistance to the anthelmintic (dewormer medication) used. This is money well spent. A parasite problem reduces productivity of your milking herd, requires milk withdrawal time for dewormer use, and costs money and time for treatment. Prevention is the best policy.

FAMACHA

The FAMACHA system was developed in Africa to identify heavily parasitized goats and sheep. This low-cost method of selecting and treating only severely affected animals increases treatment success while decreasing expense and the pace of dewormer resistance. Some parasites, such as the barberpole worm (*Haemonchus contortus*), feed on the host's blood, thus producing anemic conditions. The goat keeper, using a five-color chart on a card, compares the numbered colors—1 is red, not anemic, 5 is pale, very anemic—to that of the goat's inner eyelid membrane (conjunctiva), and records the results. Treatment is based on these results, an individual animal's past records, and herd history.

In the United States, the FAMACHA system, in conjunction with fecal egg counts, can help goat keepers reduce levels of parasites in their animals by identifying and properly treating goats with high parasite loads.

Signs of Parasitism
1.   Unthriftiness/thin
2.   Rough hair coat
3.   Soft to watery feces
4.   Pale membranes in inner eyelids
5.   General malaise

Participants in the FAMACHA system must attend an educational workshop led by a certified instructor. Visit the American Consortium for Small Ruminant Parasite Control at www.acsrpc.org for additional and up-to-date information.

Also see the Body Condition sidebar on page 48.

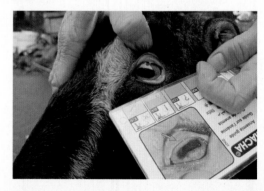

Assessing parasites using the FAMACHA method.

Begin prevention before you buy. Ask the seller for information on herd health, especially in regards to parasite resistance. Submit a fecal sample to a reliable laboratory for a quantitative fecal count with results in eggs per gram (EPG) and a hematocrit reading (red blood cell count). The farm's veterinarian should be able to assist with both of these tests with minimal expense, and they are well worth the costs. Bringing resistant parasites to your animals and soil will cost *a lot* more in the end. Always quarantine any new animals for a minimum of 30 days. Consider another fecal test before mixing the new animal with your herd, depending on the reliability of prior test results and herd history.

Prevent parasite load in your herd. Use rotational feeding areas. Feed animals up off the ground (hay, grain, water). Clean the pen regularly, especially in warm and humid weather, which is ideal for internal parasite and fly reproduction. Keep your stocking rate in check with space available for bedding, eating and browsing. Too many animals will simply create problems. Follow your plan and sell animals that don't fit. It's easy to say this one is cute and that one is so friendly. Resist this urge and take proper care of your current herd.

Feed plants that will help with parasite control. Research shows that plants high in condensed tannins, such as willows, black locusts, and leucaenia, can help control parasite infestation. Check with local extension agents for plants that will grow in your area, and be careful not to plant invasive species. Many of the effective plants are shrubs or trees, which have the added benefit of keeping goats browsing up rather than down on the ground, as they naturally prefer to reach for plants. While they vary with their environmental conditions, parasite larvae in general will only travel about 2 to 3 inches up plants, so keep goats feeding in field and wooded areas that have plant growth higher than 4 inches to further decrease the risk of parasite infestation.

Keep your goats' immune systems and health strong. Make sure they are getting a variety of forage, salt, and mineral supplements—especially during winter feeding—and water. Sound nutrition, exercise, airflow, and a calm environment keep animals strong.

Several diagnostic tools are available for determining parasite loads, from simple and inexpensive to complex and costly.

FAMACHA – What has been popular now for many years in the United States started out as a low-cost tool for sheep farmers in Africa and has become a mainstay for parasite assessment around the world. Inquire with your veterinarian, local Extension, 4-H, or goat club for workshops teaching FAMACHA. I recommend that you attend a class and learn how to use the technique, obtain a card, and schedule routine examinations of your animals. Conduct body condition scoring at the same time. Record the results, and observe changes according to the seasonal stressors.

Hematocrit – This test basically looks for the volume of red blood cells in a blood sample. A low result indicates anemia. An excessive population of *Haemonchocus contortis*, barberpole worm, drains blood from the stomach of the goat, causing anemia. There are other causes of anemia, such as blood loss injury or chronic disease, which should be considered.

Fecal sample – A fecal sample sent to a diagnostic laboratory will tell you the type and population of parasites in the goat's intestinal tract. The quantitative fecal sample is a count of parasite eggs being shed from the adult parasite population in your goat, reported as eggs per gram (EPG). Parasites are quite clever. Changes in the goat's immune system influence the animal's parasite burden. Parasites can increase egg production during times of animal stress, such as kidding, illness, and poor body condition. This increase corresponds with higher chances of parasite eggs survival.

There are currently three main classes of anthelmintics: Avermectin (Ivomec, Moxidectin), Benzimidazoles (Safe-Guard, Valbazen), and Imidazothiazoles (Levamizole). Some goat keepers experience good success with herbal dewormers. Monitor your animals with the methods above to determine the parasite burden in your herd and in individual animals. With the current variability of parasite vulnerability to treatments, I recommend working with your veterinarian, goat mentor, or local extension agent to develop a successful parasite control program.

Goat parasite identification, diagnosis, prevention, and treatment are active research subjects around the world. Information changes and updates frequently. Stay informed, monitor your animals, and be diligent with testing before bringing new animals to your property.

Check resource page for more and current information.

2. Respiratory/Lung Problems

Pneumonia and lung congestion can have many causes, including bacteria, viruses, fungus, and parasites. Identification of the organism(s) can be challenging and expensive. Generally, treatments are based on historical health information and experience.

Keep your goat-housing area well ventilated and draft free. Avoid moldy or dusty feeds. If goats are kept on a dry lot, an occasional water sprinkling may be necessary to keep dust down. Goats browsing amongst grasses and plants that are shedding pollen may react to this irritant/allergen.

Goats being transported are particularly susceptible to respiratory issues due to the stress of movement decreasing their immune system functions. A preventative course of antibiotics may be beneficial depending on the circumstances. Ask your veterinarian for recommendations. Along the same line, *always* quarantine any animals being moved *to* your herd. Thirty days is ideal, with an absolute minimum of 14 days. The quarantine pen should *at least* be far enough away to avoid nose to nose contact. A separate building is ideal. Do not risk the health of your current herd with poor planning.

3. Bloat

Fermentation of feed within the rumen naturally produces gas, which is normally released through burping or eructation. Bloat is a buildup of unreleased gas in the rumen. There are two types of bloat: primary or frothy, where the gas forms in a stable foam mixed with rumen content, and secondary or free gas. The frothy bloat is generally caused by a sudden change of feed, such as turning your animals out onto a lush spring pasture or goats finding free access to the grain bin and overeating. Secondary bloat is often caused by a blockage in the esophagus or throat so the gas cannot escape. The gas pressure buildup affects blood flow and blood pressure, and heart and respiratory rates. The pressure must be relieved swiftly or death will occur.

Learn to diagnose bloat and have treatments available to quickly relieve the symptoms. First, check for blockage in the mouth and throat. A visual and hand scan of the esophageal area can help determine if there is a physical blockage. Antifoaming bloat release drenches work quickly. A common treatment uses a surfactant, docusate, to break up the foamy bubbles and allow the gas to coalesce and be released. Follow the manufacturer's recommended dosage. Diluted laundry or dish detergents also work for breaking up the trapped gas. Passing a stomach tube into the rumen can help relieve the secondary free gas form of bloat by removing or bypassing physical obstructions. Massage the rumen and walk the goat if you are waiting for help to arrive. Be prepared to treat bloat with proper remedies and knowledge. Ask your goat mentor or veterinarian to help you learn about this easily preventable and treatable ailment before you needlessly lose an animal. (See photos on page 131.)

4. Foot Rot

Foot rot is a preventable condition that can be costly and difficult to eliminate once established in a herd. This disease is caused by two anaerobic (thrives under conditions with no oxygen), coexisting bacteria, *Fusobacterium necrophorum and Dichelobacter nodosus. F. necrophorum* naturally exists in the goat's large intestine and is generally found in soil and manure in yards and pastures. *D. nodosus* is more sensitive, surviving in soil for only 10 to 14 days. Cool, wet conditions favor these organisms, especially in areas of accumulated mud and manure. First, the goat's tender inner hoof area becomes irritated, breaking the skin barrier. These bacteria enter the hoof, and under favorable conditions, break down the hard hoof wall; the soft inner tissue becomes inflamed. There are varying degrees of severity and corresponding treatments, including a virulent form that requires aggressive treatment.

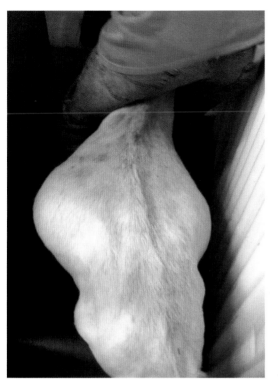
Bloat is easily seen in unhealthy goats.

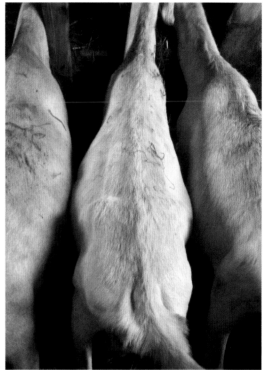
Relaxed rumen following effective bloat treatment

Again, prevention is the best management. Isolate new animals for 30 days. Avoid accumulation of wetness, mud, and manure in goat areas. Keep an active hoof trimming and foot care schedule. Difficult infections can be treated by soaking feet in a 10 percent zinc sulfate or copper sulfate bath for 15 minutes every 5 to 7 days. The entire herd needs to be soaked, with the best success occurring after hoof trimming to expose the affected areas. Remove environmental conditions that proliferate the bacteria, and fence animals away from this area for at least 14 days.

5. Urinary Obstruction
Urinary obstruction generally occurs in male goats, and particularly castrated pet goats. As mentioned, crystals of calcium phosphate and magnesium ammonium phosphate form in the urine, bind into stones (calculi), and block the outflow of urine through the urethra. The nature of the problem is related to nutrition, anatomy, and genetics. Too much grain, improper balance of calcium and phosphorus in the diet, minerals in the water, and other feed factors can predispose goats to the formation of urinary crystals and calculi. Adequate or increased water intake can help flush the urinary system and minimize aggregates of the crystals. Males castrated at a young age may have a narrowed urethra due to less testosterone circulating in their bodies. Calculi or crystal buildup is more likely to get stuck in this narrower channel. There are a variety of recommendations about when to castrate a male kid. Some think male kids should be castrated at a young age to minimize the stress on the animal, with the risk of a narrowed urethra making them more susceptible to urinary calculi blockage. Others feel, with proper technique and pain relief, castrating at 4 to 8 weeks is better for the animal. Some genetic lines are also more prone to the condition.

The formation of calculi can be minimized by proper, balanced feeding, especially of neutered pet goats. Refrain from giving extra grain and treats to "Billy," and teach children and visitors to respect this restriction. Provide dry hay, even for bucks and wethers on pasture, and provide plenty of water. Be sure salt is available, as this encourages goats to drink, especially in cold weather.

Ammonium chloride, commonly found at agricultural stores, can be added to water to prevent the development of crystals. Follow recommended dosages and limit to use with males.

6. Plant Toxicities

Some plants at various stages of their growth can contain chemical compounds that are poisonous to goats. Luckily, goats are browsers that eat a little of this and a little of that. Given access to plenty of forage, goats will generally only eat a small amount of a possibly toxic plant, minimizing the effects. Symptoms of plant poisoning include vomiting, painful bleating, kicking or biting at sides of the body, twitching, or convulsions. Symptoms vary by plant vary. I highly recommend you learn about plants in your locale that can be problems for your dairy goats. The prevalent poisonous plants in our area in the Northeast United States are members of the rhododendron family, yews, and wilted cherry. Years ago I had a group of growing bucklings penned in a fence row area cleaning up late summer growth. A surprise frost hit overnight, wilting the wild cherries. Cyanide concentrates in the *wilted* leaves of cherry—fresh and dried leaves are no problem. Cyanide kills quickly and I had six dead bucklings in the morning. What a tough lesson.

Allow plenty of plant growth before putting goats into woods or fields in the spring, and be careful of early frosts in the fall! If practical, remove plants that may be toxic or browse goats in areas primarily free of toxic plants.

Remedies vary depending on the severity of the poisoning and the time delay, as well as the individual animals. Activated charcoal products, such as Toxiban, are popular for treating toxicities. Follow manufacturer's recommendations for dose, and when in doubt, consult with your veterinarian or goat mentor before giving any medication or remedy.

7. Neurologic Diseases

Listeriosis, polioencephalomalacia (PEM), and meningeal worm infections are some of the neurologic diseases seen in goats.

Goats are sensitive to *Listeria ssp.,* a common environmental bacterium that proliferates with favorable conditions. A goat will generally succumb to this organism without immediate treatment. The *Listeria* organism causes small vacuoles in the brain to disrupt the nervous system signals. At an early stage this causes the appearance of paralysis of one side of the body: an ear will droop, one side of mouth and jaw will slack. The animal will often walk in a small circle, hence the laymen's term, circling disease. Immediate treatment with Penicillin will generally alleviate the symptoms, and the goat will be back to normal. You must determine the origin of the listeria infection and remove the animals or the source. The most common source is poorly cured silage or moldy hay. Feed silage with scrutiny.

Remember to discard milk from dairy animals treated with penicillin and other medicines. I recommend you submit a milk sample to a local dairy lab or cheese maker to test for antibiotic residue. Your veterinarian can also help with this. The residue withdrawal times listed for antibiotics do not generally apply to goats, because goats metabolize medicines at different rates than other farm animals. Test the milk from any animal that has been treated before drinking or using it.

PEM occurs in a goat that has a thiamine deficiency. This deficiency may be linked to an excessive production of thiaminases (thiamine enzymes) by rumen bacteria. This thiaminase increase can be seen in animals eating high grain rations, having sudden dietary changes, or being administered anthelmintics (dewormers). Signs are a classic "star gazing" with head tilted backward, severe arching of back, muscle tremors, and teeth grinding. Recovery from the condition depends on severity and timing of diagnosis and treatment. Administer thiamine (intramuscularly or intravenously) under advice of a veterinary professional or goat mentor.

8. Mastitis

Mastitis is inflammation and infection of the udder, or mammary gland, typically caused by a bacterial infection. Mastitis is caused by the overgrowth of an unfavorable organism in this nutrient-rich gland. Milk, especially at the body temperature of around 102°F (38°C), is a perfect medium for bacteria and yeasts to proliferate. Dairy goats are generally cleaner animals than dairy cows, due to their dry manure and clean habits. Their fastidious nature and general ease of milking make goats less prone to mastitis than cows. The downside is that goats tend to not show symptoms of mastitis as early and noticeably as cows. Many cases of mastitis in goats are subclinical, meaning the infectious organisms are present in the milk but below a level that causes visual and physical symptoms, thus allowing cases of mastitis to go undiagnosed and untreated. Physical symptoms include a clotting of the milk, swollen and hard udder, and sometimes a fever. Over time the animal's immune system may effectively decrease the bacteria load, or with an end-of-lactation dry-off, the organisms may perish without milk present for growth. The other side of the story is that the underlying growth of mastitis bacteria can proliferate during times of stress, such as a parasite load, summer heat, or other physical disruptions. Mastitis organisms include *Streptococci* spp.; *Staphylococci* spp., coliform spp.; and *Arcanobacterium pyogenes*.

You can submit milk from dairy goats suspected of having mastitis to a mastitis and microbiology lab for evaluation. Contact a local extension or state department of agriculture agent for a mastitis lab in your area. A commercial herd, especially one raised for selling raw goat milk, will need to test regularly for mastitis bacteria. Samples sent in for evaluation can be tested for sensitivity to available types of antibiotics so the most effective type can be used. These medications may be administered in the mammary gland (intramammary) and also systemically (intramuscular, intravenous, or subcutaneous), for greater efficacy. Milk withdrawal times need to be considered when determining antibiotic use. While various treatments and remedies are available, complete eradication of mastitis-causing organisms can be difficult. Sometimes culling is the only solution to a mastitis case.

Prevention is the best way to manage mastitis. The following help prevent infection.
- Keeping goats healthy and well-fed, with access to well-managed pasture
- Keeping housing, bedding areas, and exercise yards dry and clean
- Washing hands before and after milking
- Using clean, well-kept milking equipment
- Cleaning teats pre-milking and dipping them post-milking
- Building strong immune systems
- Monitoring overall animal health
- Preventing older kids from nursing to avoid increased risk of introducing environmental organisms into the udder

9. Breeding Difficulties

Problems with breeding are generally caused by unbalanced nutrition, uterine infections, or physical difficulties such as sore feet and legs. All goats, both females and males, need to have a sound feeding program, especially for 3 to 6 weeks prior to breeding.

Does are most likely still milking when being bred and need proper nutrition to meet all their demands. Unbalanced feeding can lead to unbalanced hormone levels, which lead to infertility and estrous cycling problems. There are commercially available hormone treatments with such ingredients as GnRH (gonadotropin-releasing hormone), prostaglandin, and progesterone that can help jumpstart or short-cycle goats that are having difficulties. Having a balanced diet with sufficient energy, protein, vitamins, minerals, and water is important. Goats that are on a stressed pasture during a hot summer may have difficulties with fertility, especially if there is also a parasite overload. Be sure female kids get bred *only* when they reach a proper weight. For full-size dairy goats, I want kids to be 75 or 80 pounds when they are bred. Our doelings generally reach this weight when they are 8 months old. This age brings them into birthing at age 13 months, which is after the adult goats have given birth. Following this schedule allows more individual care to the younger animals that are birthing for a first time and need special attention.

Bucks also need to be in prime condition for breeding. Be sure to increase their plane of nutrition several weeks prior to breeding season, but avoid over-conditioning (overweight). Strong and sound feet and legs are critical for a breeding buck.

10. Nutritional and Metabolic Diseases

Nutritional and metabolic issues in goats can be the most difficult to diagnose and treat. As mentioned at the beginning of this chapter, goats often do not show signs of these deficiencies until it's too late to correct. Pregnancy toxemia (hypoglycemia, or low blood sugar) and milk fever (hypocalcemia, or low blood calcium) are generally seen 1 week prior to 8 to 10 weeks post-kidding. Grass tetany (hypomagnesemia), white muscle disease (vitamin E and selenium deficiency), and polioencephalomalacia (thiamine deficiency) also fall into this category. These diseases can be corrected with accurate diagnosis and prompt action. Learn to recognize the slightest changes in your goats when conditions are reasonably possible for these diseases to occur.

Pregnancy toxemia, also known as ketosis, commonly occurs in the last week of pregnancy, or even at birthing. This illness often occurs in goats that are pregnant with multiple fetuses, over-conditioned (fat), or timid at feeding time. Basically, the doe's body isn't able to keep up with the energy demand at late pregnancy. The liver kicks in to help with the demand by metabolizing fat stores. Too much demand overwhelms the liver and its ability to function properly. Prevention is important. Keep goats in good condition year round. Goats that are overweight at breeding will be overweight at kidding. Encourage shy goats to eat with the herd, or feed them separately. Promote animal movement and exercise throughout the pregnancy.

This particular condition requires careful attention to the slightest change in a late pregnancy doe. Early symptoms include being off feed, hesitation to move with the herd, and having a general lethargy. This imbalance quickly spirals downward without increased glucose or decreased demand (birthing to remove fetal energy drain). The accumulation of ketones (a side effect) will produce a sweet, rose-like scent on the breath and can be detected in the urine. Goats with this symptom are heading for trouble and need attention immediately. For early treatment, oral administration of propylene glycol (60 ml twice daily) for 2 to 3 days can help balance the demand. Administering insulin and dextrose may be necessary, also. Signs of advanced stages, such as incoordination, weakness, inability to stand, and blindness, require immediate veterinary assistance with administration of intravenous solutions and fluids. **Do not wait to treat this condition. A very pregnant doe will quickly move past a point of treatment and die.**

Milk fever, or hypocalcemia, is an imbalance that generally occurs in goats 6 to 8 weeks after kidding. This condition often occurs in animals with high milk output. Goats are able to absorb nutrients efficiently in their systems and delay symptoms of this deficiency until weeks after birthing. Calcium and phosphorus work together, and the feed ratio of 2:1, Ca:P, is important to maintain, especially during pregnancy. Feeding hay high in phosphorus and low in calcium during the dry period (late pregnancy) may throw off the ratio enough to cause this condition once the doe has given birth and the body's demand for calcium increases with lactation onset.

Classic signs for less severe cases are poor appetite and listlessness a few weeks after kidding. This mild deficiency can be corrected with oral calcium supplements or calcium-containing antacids. More severe cases show signs such as weakness, inability to stand, twitching, and ears laying back. Afflicted goats can also be hyperactive. Pupils are dilated and react slowly or not at all to a light shone at them. These cases need to be treated quickly with careful intravenous administration of calcium borogluconate solution. Seek veterinary care for this treatment as improper administration can quickly kill your goat.

Response to treatment of severe cases is dramatic, with return to normal within a short time. Monitor this goat closely for the next several days, as an occasional relapse can occur.

Plants such as raspberries, garlic, mint, and parsley can be made into tea and administered to goats.

Complementary Health Care

Complementary health care, such as herbal teas and ointments, and homeopathic remedies, can balance a goat's well-being. Learn about herbal remedies from a local herbalist, resources listed in back, and simply offering safe herbs to your goats. Mints; comfrey; cooking herbs such as dill, oregano, thyme, basil, rosemary, and sage; field plants such as chickweed, nettles, rose, raspberry, and pine bough all contain minerals and essences that are beneficial to goats. A particularly useful resource is the classic *Herbal Handbook for Farm and Stable*, by Juliette de Baïracli Levy. Keep notes on herbal remedies you try and use them to create your own herbal chart. *Educate yourself on the plants you are using, as some plant parts or growth stages can be toxic.* Use knowledge and common sense.

Economics of Health

Determining how much money to spend on healing an animal is not easy. No one wants to lose an animal to a treatable illness. Decide on a budget for veterinary care and try to follow this with considerations. Backyard goat keepers and specialty goat breeders will most likely cushion their veterinary budget a bit more than a commercial goat farmer. This in no way means there is less care or concern for an animal. There may simply be less cash flow. As expressed before, goats should not be a financial drain on your family income. Use the information in this chapter, the resources listed in the back, your own experience, goat/animal forums and groups, your mentor, and a veterinary professional to assess an unhealthy goat and *act* on that information.

Veterinary Homeopathy, by Laurie Loften, D.V.M., C.V.H.

Homeopathy is a medical treatment modality developed more than 200 years ago by Dr. Samuel Hahnemann, a German physician. Hahnemann discovered that illness affects the whole patient: mentally, emotionally, and physically; and that each patient has his or her own unique responses to disease, or disorder, in the body. Thus the treatment, in order to be truly curative, must take into account the totality of signs experienced by each patient. Conventional medicine takes the approach of treating the disease itself, not the whole patient (or the underlying cause of that disease). This form of treatment often shifts the signs to another body system, masking and suppressing the signs, and leaving the patient still feeling unwell despite apparent resolution of the primary complaints. The basic premise of homeopathy is that there is a single homeopathic medicine (chosen from hundreds) that addresses the totality of the patient's signs at the time, stimulating a curative response and allowing the disease to resolve as the patient recovers. Thus care must be made to observe *all* signs: mental, emotional, and physical, to then determine which medicine is most similar (homeopathic), and thus curative, to the patient.

Homeopathic medicines are made from mineral, plant, and animal substances. They are prepared in a rigorous manner to create different strengths, called potencies. The lower potencies are available without prescription in health food stores. The challenge for the layperson is that many choose a homeopathic medicine based on the single indication, such as "diarrhea," listed on the vial. Hundreds of homeopathic medicines address diarrhea, but the one selected may not be the correct match for the whole patient. Thus it may seem that the medicine is "not working." This is because it must fit the whole case, all symptoms of the patient. When the right medicine is chosen, based on the totality, the resolution of clinical signs can be smooth, gentle, and free of complications. Choice of potency depends on the age, strength, and wellness of the patient. As a beginner, and if you are treating on your own without the guidance of a homeopath, the 30cc potency is frequently a useful and safe choice. There is no meat or milk withholding time, another benefit in farm animal practice.

Homeopathic medicines and indications frequently helpful for goats:

- *Arnica montana* – Trauma, including post-partum; fears touch or approach of anyone; pains insufferable with restlessness; can't get comfortable.
- *Arsenicum album* – Exhaustion, prostration with restlessness, worse at night; unquenchable thirst; diarrhea; cadaverous odors; burning pains.
- *Belladonna* – Fever; inflammation; mastitis; pregnancy toxemia; plant toxemia; dilated pupils; delirious; twitching, jerking, seizures; redness of parts.
- *Calcarea carbonica* – Slow/poor development, tardy teething, late walking; hypocalcemia of pregnancy; weak kids; defective nutrition of glands, bones, skin.

- *Carbo vegetabilis* – Lifeless, cold breath.
- *Hypericum* – Injuries to nerves, especially extremities; excessive painfulness.
- *Nux vomica* – Irritable, doesn't want to be touched; digestive disturbances; toxicity.
- *Pulsatilla* – Changeable symptoms, clingy, shortness of breath, thirstless, malpositions of fetus, labor pains weak/ceasing, false pregnancies, poor milk letdown.
- *Sepia* – Indifference/aversion/anger toward own kids, won't let them nurse.
- *Silicea* – Timid, overly sensitive, startling, abscesses; treatment promotes discharge of pus.

Thousands of remedies are discussed in Materia Medicas (encyclopedias describing homeopathic medicines). Acute and chronic problems, as well as several kinds of injuries, can be treated. Chronic illness treatment and resolution is more complicated and should be managed with the aid of a homeopathic practitioner. Along with homeopathic treatment, it is still necessary to provide supportive care, appropriate shelter, nutritious feed, and water.

Death

Death is the end of life, birth the beginning. This handbook has an entire chapter about birth, yet only a few paragraphs about death. Funny thing isn't it?

Growing up on a farm taught me the lesson of life and death at an early age. My siblings and I all have our stories—a cat, a cow, a dog, a chicken. My first personal hardship experience with animal death came when I was around eleven. We had a dairy calf that would not suckle. This particular calf didn't nurse off the mom and wouldn't take milk from us either. Everyone tried. I usually had good luck getting calves to drink, so this became my job.

I was doing all the tricks that usually worked: warm milk, tickle the nose, hold the chin, stroke the neck and ribs, suckle on the finger, and more. This calf just wasn't right. A day went by. She was getting weak and wouldn't even stand. To this day, I remember those big calf eyes looking at me. She wanted to drink, but something wasn't right. As I was trying again to get her to suckle, she looked hard into my eyes, and then with a sweep, I saw life leave hers. She was gone, lifeless, in my scrawny arms. As I held her, I promised that calf that I would not let an animal pass without trying my best to find a remedy or at least reduce or relieve its suffering.

To this day, I have honored that promise. When you spend your life with sixty-plus farm animals, you witness plenty of deaths after many years. I do wish it would get easier, but the plain fact for me is that it gets harder. We all have our measures. I don't have an endless veterinary expense account. I have to be reasonable. But my learning is endless, and that is really what counts. I'm sure I cried when that calf died. The family was busy milking, so it was just the two of us. My gut told me to give thanks and respect. A life lived, if but a few days, and a promise made that continues today.

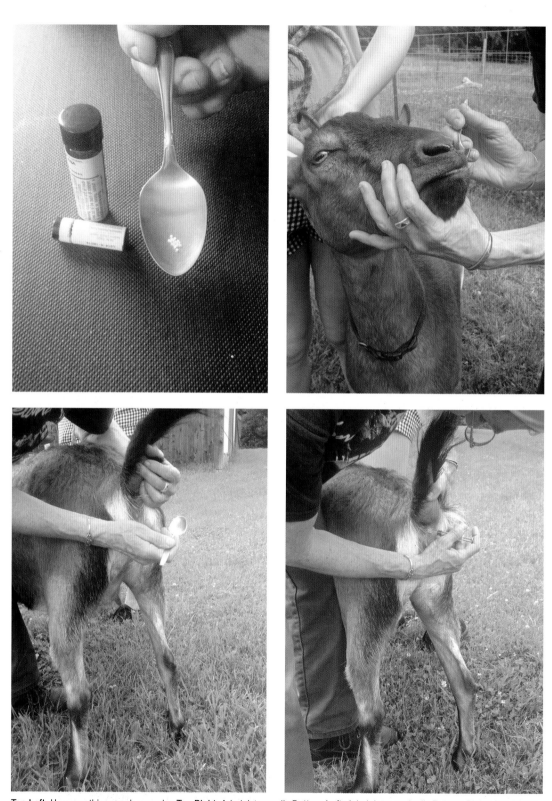

Top Left: Homeopathic remedy granules **Top Right:** Administer orally **Bottom Left:** Administer vaginally **Bottom Right:** Administer vaginally after diluting in water. *Lita Lofton*

Cherish the beginnings, respect the endings, and savor the in-betweens. Concentrate on the wellness of your animals. Be aware when problems occur; intervene to help fight off sickness and heal your animals. Learn from the goats, take notes, and use resources available to you. Keep your animals' immune systems strong and their whole bodies well balanced. Give them nutritious feed, fresh water, exercise, clean housing, and plenty of care.

"One thing to remember is to talk to the animals. If you do, they will talk back to you. But if you can't talk to the animals, they won't talk back to you, then you won't understand, and when you don't understand, you will fear and when you fear you will destroy the animals, and if you destroy the animals, you will destroy yourself."

—Chief Dan George

Some goats have an almost preternatural sense to understand your needs or language. Talk to them often!

Chapter 10

Milk and Milking

Goat milk, in its simplest liquid form, is a nutritious and flavorful food. Milk is a beneficial source of protein, fat, carbohydrates, minerals, and vitamins. Goat milk is universally considered an acceptable nutrition source for many species of orphaned animals.

The major components of goat milk are water, protein, fat, and lactose (milk sugar). It also contains minerals, vitamins, and enzymes. The water fraction in the milk is around 87 percent. The protein component is primarily casein, with lactalbumin and lactoglobulin also present. Protein accounts for around 3.5 percent of the milk's makeup. Casein is an interesting, complex, and still not yet understood matrix of protein, calcium, and phosphorus. The lactalbumin portion of goat milk is structurally different than that of cow milk. Some people who can't tolerate or are allergic to cow milk protein can drink goat milk with no reaction.

Fats average 4.0 percent in milk. Triglycerides make up around 95 percent of goat milk fats, with higher levels of caproic, caprylic, and capric fatty acids compared to cow milk. These specific fatty acids give goat milk and cheese its full flavor. The goat milk fat globules are slightly smaller than fat found in cow milk. Goat milk is known for being naturally homogenized, meaning the fat does not clump together and rise to the top of the milk. This is because the protein lactadherin, which is part of the fat globules of the milk, has a different structure in goat milk than in cow milk. The type of lactadherin in goat milk does not cause fat globules to clump together and form a cream line.

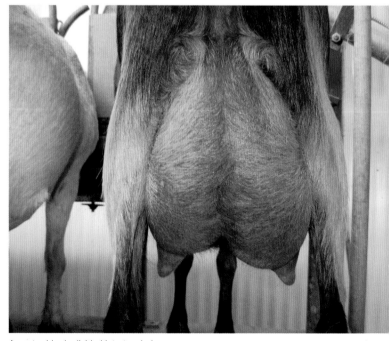

A goat udder is divided into two halves.

Lactose, or milk sugar, is a good source of energy and an important component of the milk for its conversion in dairy products. Lactic bacteria convert lactose to lactic acid, which in turn coagulates the protein components. Goat milk is composed with about 4.6 percent lactose, which is only slightly less than the percentage in cow milk.

Most major minerals and vitamins are found in goat milk, which is also a good source of calcium, phosphorus, potassium, sodium, and magnesium, especially if the goats are able to browse and eat fresh forage. Part of their selective feeding habits is to procure a range of nutrients, and particularly trace minerals. Goat milk offers many vitamins, particularly A, thiamine, and niacin. Goat milk lacks the vitamin A precursor, carotene, which gives the yellow color to milk from some cow breeds, making goat milk and milk products pure white.

Milk is secreted in the mammary gland, or udder. The udder of a dairy goat is separated into two halves, with a single gland in each half.

These halves are walled compartments, so milk does not flow from one half to the other. Each half has one teat at the bottom of the gland where milk is released from the body. The halves are divided and supported by medial suspensory ligaments. The udder suspensory ligament system and body attachments are important for keeping the udder in place, especially for all the weight this gland holds. A goat's udder can carry ten to twenty pounds of weight at any time. Walk with twelve pounds of sugar in a bag for a while and see how your arm feels. In addition, this support system has an elasticity that allows for expansion of the udder with the accumulation of milk.

Inside the udder are very small secretory cells called alveoli. This is where the magic happens. An alveolus has a diameter of 0.01 to 0.03 millimeters. Despite the tiny size, their function is huge. These secretory cells take blood nutrients from surrounding capillaries and make milk. Sounds simple, until you stop to think that the blood flowing through these tiny capillaries contains the building blocks that become milk. This milk is different in

Milking your goats regularly is an important habit for a happy and healthy herd. *Shutterstock*

composition and structure from the originating blood. The small alveoli convert the blood nutrients into sweet, healthy milk. Just amazing! Impressively, around 400 units of blood are passed through the udder for every unit of milk produced. Consider how many times a goat heart beats to push blood through the udder to make a quart of milk. Goats are certainly working 24 hours a day. They deserve a rest under the shade tree.

The manner in which goat milk is secreted is called apocrine secretion. Interestingly, cows secrete milk through a different cellular mechanism. Basically, with apocrine secretion, the milk particles ascend to the high point of the epithelial cell and are pinched off with a portion of the cell membrane and some of the cell cytoplasm. This castoff cellular material can cause problems with a milk-quality test called Somatic Cell Count (more on this later).

Alveoli are clustered together in grape-like lobules that form together into lobes. The lobes are connected in a series of ducts to finally gather milk in the udder cistern. This cistern can only hold about a pint of milk, while the remainder of produced milk is held in the ducts, lobes, and alveoli. The alveoli slow milk secretion as mammary gland pressure increases with the accumulation of milk. Milking at a regular interval, such as every 12 hours, is important for the release of gland pressure and also to continue milk production. Milk accumulated and unreleased for a longer period of time can cause mastitis (infection) and other health issues. Primarily, though, the built up pressure will signal alveoli to stop secreting milk. The accumulated milk can resorb with time, and the goat will cease milk production until she births again.

The udder cistern is connected to a small teat cistern (chamber inside the teat). The milk is held in place through a series of intricately folded mucous membranes, called Fürstenberg's rosette, and a circling of sphincter-like smooth muscle fibers. The Fürstenberg's rosette is located at the internal end of the streak canal. This special structure retains milk in the cistern and is also important in preventing bacterial movement into the gland. The streak canal is a short, tube-like structure located between the bottom of the teat cistern and the teat orifice (opening). Its keratin-based lining has anti-bacterial properties. The milk gathered in the teat cistern has the highest bacterial levels and should be discarded. Strip this milk out first to feed to your dog, barn cats, or pigs.

Through various stimuli at milking time, the goat releases oxytocin, which causes muscle contraction. The muscle cells surrounding the alveoli, ducts, and cisterns contract, sending milk down the series of channels and cisterns and out the teat orifice. The retaining membranes relax enough to allow milk to be removed from the udder. Milk secretion reconvenes quickly when the accumulated milk is taken from the udder. With the completion of milking, oxytocin flow stops, muscles relax, retaining membranes tighten, and with the stored milk pressure removed, the magical conversion of blood to milk begins again.

After years of working with dairy animals, I still stand in awe watching our dairy goats selectively browse, munch sweet-scented hay, and drink cool water, knowing all this goodness is being converted first in their gastrointestinal system to the circulatory system to the mammary system and then to my cup!

While not practical for many, the most nutritious way to drink milk is warm, from the goat to your cup to your body. I enjoy a morning cup of milk with a swirl of black strap molasses—my milk-uccino.

Somatic Cell Count (SCC)

Somatic cell count (SCC) is a milk quality test that indicates the overall health of the udder. This test counts leukocytes, or white blood cells, in the milk. Leukocytes are part of the immune response produced by an animal to aid in fighting off infection. This test was developed for cow milk. Goat milk, due to its nature of secretion, needs a special stain, as the cast off cellular material from the apocrine secretion resembles a leukocyte. This causes an artificial high cell count with the regular staining method. Goats also have a higher count during estrus and at dry off period, with no relation to increased infection. These differences are recognized nationally, and higher limits are in place for goats. Goat milk limits are now 1,500,000 cells per milliliter, while cow milk is 750,000 cells per milliliter. The Dairy Practices Council is an excellent resource to learn about goat milk quality, tests, and much more. For more information, see www.dairypc.org/catalog/guidelines/small-ruminants-goats-sheep.

My milk-uccino—delicious!

Goats gather the goodness of forage to convert into milk.

Drinking raw milk is not for everyone. The beneficial bacterial flora may take some time adjusting in your body. Try fermented dairy foods, such as yogurt or kefir, to help adjust your system. Happily, I grew up on a family dairy cow farm and have spent most all my life drinking raw milk. This wholesome drink remains one of my favorite beverages.

The raw milk question stirs a lot of debate. There are some very real dangers to drinking raw milk. Salmonella, listeriosis, Q fever, brucellosis, toxoplasmosis, and many others maladies can be passed in milk. Know the health of your animals and the cleanliness of their quarters. Keep the milking area clean and equipment clean and sanitized. Get the milk tested monthly to check bacteria levels. If you have any doubts, discard milk from suspicious animals and get these animals checked by a veterinarian or goat mentor. Do not risk illness or potential death for a cup of milk. There are many good resources on this topic. Seek information on all angles of the debate, and decide what works for you and your family. A benefit of keeping your own dairy goats is you know their health and the care that is given both to them and their milk. Inform yourself on the issue, and do not take risks with your and others' well-being.

Pasteurization is the technique used to heat milk to minimize harmful bacteria. This heating also minimizes the good bacteria in milk, as well as naturally occurring enzymes that aid in the digestion of milk. Pasteurization is required by U.S. federal law for any licensed dairy products sold that are not aged at least 60 days. The gentlest method of pasteurization is to heat the milk to $145\,°F$ ($62.8\,°C$) and hold it there for 30 minutes while the milk is constantly agitated. Official pasteurization includes temperature-monitoring steps and air-space heating. Contact your local, state, or federal health official for more information on this process.

Milking a goat can be done by hand or with a machine. I enjoy hand milking, but, with sixty goats to milk every day, twice a day, we use milking equipment. Beginners can pick up the technique quickly once they build up confidence and some hand and lower arm strength. See the video on Milking at vimeo.com/user29715351.

Dairy Herd Improvement Association (DHIA)

The Dairy Herd Improvement (DHI) program helps dairy farmers across the country improve their business and animals' health and production by gathering and organizing milk production information. Milk samples are collected monthly and analyzed in approved laboratories, with farmers receiving detailed information on milk quantity, quality, and additional data for dairy animal management. This information is important for both individual animals and herd improvement. Make decisions on animal health, breeding, culling, nutritional needs, milking equipment maintenance, and many other aspects of keeping a herd of dairy animals based on this information. While started for dairy cows, most areas offer opportunities for goat farmers to participate as well. Smaller producers or backyard enthusiasts may be able to join together to submit samples. Contact the service provider in your area and discuss how a program can be structured. For more information, see www.dhia.org's resource tab for service providers as well as DHI programs throughout the country.

Procedure for Hand Milking

Equipment
Sanitizer solution (bleach, water)
Small bowl of soapy water and clean towels
Clean, sanitized, and drained milk bucket
(stainless steel is best)
Lidded strip cup
Funnel
Filter or strainer
Iodine-based dip

Preparation
1. Wash your hands.
2. Sanitize milking equipment using 2 tablespoons of bleach for 4 quarts of water. Always add the bleach to the water and mix gently without splashing.
3. Make a small bowl of warm, soapy water using a non-irritating soap—homemade goat milk works well!
4. Set up the milk stand, putting grains in a feed bucket if desired, and retrieve your animal.
5. Gently wash the goat's udder with the soapy water and a damp cloth. Do not use a lot of water or make a mess. Goat's udders generally stay very clean if they have a clean pen and paddock area. You should simply be wiping away dust, hair, and other materials that may fall in the milk. Make sure the teat orifice (opening) is properly cleaned and free of dried debris. If the goat's udder is caked with mud or manure, you need to clean the pen and do a better job with management.
6. Dry the udder with a clean, dry cloth or paper towel. You are ready to milk.

Milking Method
1. Wrap your thumb and forefinger around the base of the teat (where it joins with the udder). You don't want to grab the udder, just wrap at the top of the teat.
2. Press your thumb and finger together, with the teat in between. *Hold this tightly*

Hand-milking technique hasn't changed in eons. Don't be shy! *Jay Iversen*

Strip cup used to check initial milk for problems. Discard this milk. *Jay Iversen*

with even pressure so milk doesn't move back up into the udder while you firmly roll your other fingers down the teat toward the palm of your hand, making sure to move your fingers from the top down. Milk from the teat should excrete easily with a little pressure.

3. When this milk is squeezed out, open your hand and thumb to allow the teat cistern to fill again with milk. Repeat.

4. The first several squirts of milk are from the teat cistern and should be discarded. Strip it into a lidded cup and observe for mastitis clumps, blood, or other problems.

5. Using your clean and sanitized milk bucket, milk until all milk is comfortably removed from both udder halves. With some practice, you will quickly be milking both halves alternately, like the pros. Some goats will have a second letdown. Removing milk from this second letdown is optional. You can take any milk that refills the udder within a few minutes, or allow it to remain. There is no harm in allowing some milk to remain in the udder, although you may get less overall production.

6. When finished, dip the teat end with an iodine-based or similarly disinfectant dip to block bacteria from entering the teat while the sphincter-like structures close tightly.

7. Return the goat to her pen and retrieve another.

8. When finished milking, strain the milk through a filter. If using a funnel and round filter, fold as shown.

9. Refrigerate milk immediately for future use.

10. Clean all milking equipment by first rinsing it in warm water to remove milk residue. Wash it in hot, soapy water. Use of a dairy soap is helpful to remove milk fat and minerals. Wear gloves to protect your skin. Rinse in clear water, followed by a dairy acid (phosphoric acid) rinse to keep mineral deposits from forming on your equipment. Store equipment so water can drain from it. Sanitize just before the next milking.

11. Clean the milking area, rinsing or wiping off milk from the stand, floor, and other areas. Sweep and discard debris. Cover grain bins. Restock supplies for next the milking.

When pouring milk, watch the fluid stream going into the receiving container. Adjust your pouring to minimize spills.

Machine milking equipment can be simple to quite involved.

Goat supply catalogs have good equipment and information on setting up a system. Also see *A Guide to Starting a Commercial Goat Dairy*, by Carol Delaney, and other information on the resource page in this book. Size your system properly for the number of animals you plan to milk and the milk volume. Maintaining equipment is important for efficiency, herd health, and milk quality.

Finish by protecting the teat end with a dip of an iodine-based product. *Jay Iversen*

Small pail milking machine.

Milking parlor setup.

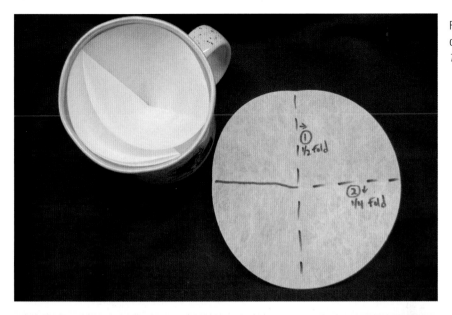

Folding a round filter disk to fit a funnel.
The Goat Dairy

Lactation Education

Dairy goats, as domesticated farm animals, give birth and produce milk for 10-plus months, go through a dry period at the end of their pregnancies, and then give birth and begin the cycle again. The quantity of milk produced during this cycle varies. Basically, the milk quantity follows the natural needs of the young offspring, which demand more milk with growth to a point of weaning. Some goats will continue milking for 20-plus months, providing year-round milk and alleviating the stress of giving birth. This extended lactation also reduces the number

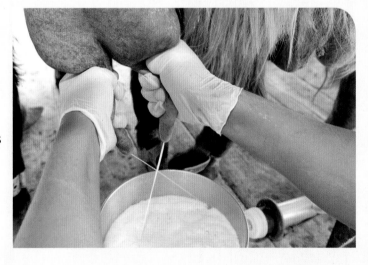

of offspring that need to be sold each year, which can be beneficial if there isn't a market for young goat kids in your area. There are many current research articles on the Internet for extended lactation in goats.

Chapter 11

Dairy Products

The nutritious and healthy milk from dairy goats can create many kinds of delicious dairy foods. Like the milk, the products you make—yogurt, kefir, cheese, butter—will pick up flavors from what the goats are eating, the season of the year, and the particular goat breed. Yogurt made from spring milk with have a light, flowery bouquet of the spring grasses.

The concentrated feeds of fall browse will bring an earthy depth and caramel note to your chèvre. Even hay harvested in various stages will bring nuances of tastes to your milk and milk products. Nigerian Dwarf and Nubian milk, with their higher milk fat and protein percentages, will yield more product per quart of milk and have a rich flavor.

Chevre is as tasty as it looks—try it with a wide variety of recipes. *Beatrice Peltre*

What fun to take the simple milk from your dairy goat and make velvety smooth kefir or melt-in-your-mouth Camembert. You can choose to pasteurize the milk first or use raw milk. As noted in the discussion on raw milk, please make sure you take milk from healthy animals, use proper hygiene when milking and storing the milk, and follow strict sanitation procedures when creating your dairy products. You have worked hard and diligently to procure this wonderful milk. Take a few minutes to properly sanitize your work space, equipment, and utensils to avoid an unsafe or failed end result. Schedule cheese- and dairy-product making for a particular day and time to avoid interruptions.

Sanitation

Bacteria—both good and bad—are everywhere. Develop a strict sanitizing process, even in your home kitchen, when working with milk. If possible, use a separate set of tools for your dairy procedures, reserving these for this task only. Errant bad bacteria, naturally in the environment, can ruin a batch of cheese, wasting your time and hard work. More importantly, pathogenic bacterial strains can cause an unsafe dairy product with potential to sicken or worse. Establish a sanitizing process and stick to it. Leave shortcuts for other parts of your day. Change into clean clothes, remove jewelry, and secure long hair. Proper sanitation is *critical*!

Wash your hands thoroughly with warm soapy water, rinse, and dry. Bring together all the equipment and tools you need for your recipe. Make a sanitizing bleach solution from 2 tablespoons of household bleach added to 4 quarts of cool water. Always add the bleach to the water, and avoid splashing bleach in your eyes or on your skin. If you can't use bleach, follow an appropriate alternative procedure for proper sanitation. Clean and sanitize your work counter and all contact surfaces. Wash your equipment with hot, soapy water, rinse thoroughly, and then rinse again in the bleach solution. Set all equipment on clean cloths or a drying surface until you are ready to use it. You can also sanitize your equipment by bringing water to a boil in your cheese pot. Place your tools in the boiling water, bring it back to a boil, and allow it to sanitize at a boil for an additional 5 minutes.

Cheesecloths need to be laundered separately from other clothes. Add perfume-free detergent and bleach to the wash and dry in a dryer (use unscented fabric softener if you use any). Alternatively, if line drying, boil cheesecloths for 10 minutes and set to cool before each use.

Follow sanitary procedures by keeping household pets and visitors out of the area during the make time. Rewash and dry your hands as needed.

The yeast from bread-making can contaminate dairy products, so plan this task for another day.

Cultured dairy foods are created by the growth of particular strains of good bacteria at the right temperature in milk. The particular bacteria and process you choose to use will create a specific product. This lactic (milk) bacterium occurs naturally in the milk.

You can simply leave milk from a healthy animal at room temperature and the milk will clabber, or set, as the lactic bacteria feeds on the lactose (milk sugar). The lactose is converted to lactic acid during this process, increasing the acidity of the milk. This increased acidity alters the milk protein strands, changing the fluid milk into a soft curd.

To make cultured dairy foods, we jumpstart the growth of beneficial bacteria by adding more of it to the milk. This larger population competes better for the available lactose, allowing faster growth of their particular strain. Other bacteria will not grow as fast and may eventually die as the desired population thrives. Imagine a soccer game with one team having five times as many players. Who will make more goals?

Bacteria

Two types of bacteria are used for making cheese: mesophilic and thermophilic. Mesophilic (mild-loving) type thrives in a mild temperature, with a range of 50 to 108°F (10 to 42.2°C). The ideal temperature range is 77 to 86°F (25 - 30°C). Mesophilic bacteria are mainly used to create the acidic environment for cheese making. Thermophilic

General Cheese Definitions

Acidity – Latin root of the word means sour, thus a measure of sourness. In simple chemistry, a measure of hydrogen ions in solution. This value is measured on a pH meter or with pH paper. pH scale goes from 1 to 14 with 7 being neutral. Lower numbers are more sour or acidic. Milk has a pH of 6.7, whey from cheese making can range in pH, generally between 4.4 to 5.3.

Curd – The solid fraction of milk after coagulation alters the milk proteins.

Lactic acid – End product of lactic bacteria utilizing lactose for energy, increasing acidity of milk, which in turn coagulates milk proteins.

Lactic bacteria – Naturally occurring, single-celled organisms in milk; also specific strains added to milk for cheese making. These

beneficial bacteria convert lactose to lactic acid, coagulating the milk proteins. Specific bacteria are used to create specific cheeses and flavor profiles.

Lactose – Naturally occurring sugar found in milk, made up of glucose and galactose. Lactic bacteria convert lactose to lactic acid.

Rennet – Enzymes that coagulate the milk protein, turning liquid milk into a solid fraction, curd, and a liquid fraction, whey. Rennet is naturally produced in the stomach lining of young goats, cows, and sheep. Rennet is also available through plant sources, such as thistles, and molds.

Whey – The liquid fraction of milk after coagulation alters the milk proteins.

(heat-loving) is the second type, thriving in a higher temperature range of 68 to 131°F (20 to 55°C). The ideal temp for this group is 95 to 105°F (35 to 40.6°C). Thermophilic bacteria are generally used for creating flavor profiles and textures in aged cheese.

The main family of mesophilic bacteria in cheese making is Lactococcus. Most starter cultures are a mix of *Lactococcus lactis ssp. lactis* and *Lactococcus lactis ssp. cremoris*. These cultures digest milk lactose, creating lactic acid and beginning the transformation of liquid milk into curd (solid portion) and whey (liquid portion). Other mesophilic cultures, including *Lactococcus lactis ssp. diacetylactis* and *Leuconostoc mesenteroides ssp. cremoris*, are used to help create aroma profiles.

The thermophilic bacteria used in cheese making is primarily *Streptococcus thermophilus, Lactobacillus helveticus*, and *Lactobacillus delbrueckii ssp. bulgaricus*. These bacteria thrive at a higher heat and are used for cheeses that are made by cooking the curds to help extract whey. They also aid in the production of lactic acid as well as other components that bring flavor and texture into particular cheeses. The firm Italian cheeses, such as Parmesan and Provolone, as well as Swiss and Gruyere, use thermophilic cultures; mozzarella does also.

Another main ingredient in cheese making is rennet, which is used to coagulate the milk proteins. Rennet contains the proteolytic enzymes, chymosin (rennin), and pepsin. Rennet breaks the protein chain at a specific place leaving portions of the otherwise protected casein molecule exposed. The exposed casein molecule can then chemically bond with other casein molecules, creating a solid mass. There are a variety of rennets: animal, plant, and microbial.

Animal rennet is naturally produced in the true stomach (abomasum) lining of a young ruminant (calf, kid, or lamb). Cheese lore says that milk was placed in a pouch made from the stomach of a young goat kid, and after a while, when the pouch-bearer took a sip of milk, the contents had firmed. The naturally occurring rennin coagulated the milk and thus the value of rennet in curd formation was realized. Animal rennet is widely used. Many consider

Cultivation of Thistle plant used for coagulation, Canary Islands.

this the standard rennet, especially for firm, aged cheeses.

Using plants to form milk curds also goes back a ways in cheese history. Many old societies used plants, mainly thistle and fig, to coagulate milk. Although vegetable rennets are not as readily available and reliable as other types, they are still used today by many traditional cheese makers, including those in the Mediterranean region.

Microbial rennet is produced by the fermentation of the organism *Rhizomucor miehei*, which is a mold grown in vats with the active enzyme extracted and purified. The microbial rennet can lead to bitterness in the cheese, especially with aged cheeses, and is therefore preferred for fresh cheeses.

In the 1980s, scientists developed a method of fermenting genetically modified fungus or yeast to produce synthetic chymosin. The chymosin is isolated and purified from the fermentation broth, with the modified organism killed. Thus the fermentation-produced chymosin (FPC) is free of modified organisms. This type of rennet is used widely in the commercial cheese-making industry, due to the stability and standardization of the product. Some smaller cheese makers also use this type of rennet, a non-animal-based product, for making kosher, halal, and vegetarian cheese. Many feel this rennet does not work well with the firmer, aged cheeses and is better suited for soft, fresh cheeses. Rennet needs to be refrigerated and kept in a dark location. Dilute it using cold Chlorine-/fluorine-free water.

Curds for cheese making are primarily formed either by lactic bacteria or rennet. Generally a combination of these two coagulants is used. Soft, unaged cheeses, such as chevre, are formed mainly by the action of lactic bacteria. They use a small amount of rennet, form soft curd over a longer period of time (18 to 24 hours), and are more perishable with a higher moisture content. Firmer, aged cheeses, such as gouda, are considered a rennet curd formation cheese. They require more rennet, coagulate quickly (20 to 60 minutes), form strong curds, and store with lower moisture content.

Note: Use only salt with no iodine, such as kosher or sea salt. Use stainless steel equipment, which is easily cleaned and does not react to the acidity of whey.

Recipes
YOGURT

Yogurt is easy to make and enjoy. Goat milk yogurt is thinner than that made with cow milk, unless you are using higher-solids milk, such as that from the Nigerian Dwarf breed. A thicker end product can be obtained by adding gelatin, skim milk powder, or a small amount of rennet. Also, the thinner yogurt can be drained in a cloth-lined colander to the desired consistency.

Equipment
- Cheese pot (heavy-bottom stainless steel pot or saucepan)
- Long-handled spoon or skimmer
- Dairy thermometer
- Small bowl or measuring cup
- Small spoon or whisk
- Measuring spoons
- Measuring cups
- Ice water in sink or large pot for cooling milk
- Large bowl or individual yogurt glasses/yogurt maker for incubation period
- Towels
- Heating pad

Ingredients
- 1 quart goat milk
- 1 teaspoon dry yogurt culture or 2 tablespoons fresh, plain, active-cultured yogurt

Optional
- ⅛ teaspoon unflavored gelatin powder **OR**
- ¼ cup skim milk powder **OR**
- 1 drop rennet diluted in 2 tablespoons cold water (dilute rennet at time of use)

Yogurt culture is thermophilic bacteria: Streptococcus thermophilus and Lactobacillus bulgaricus; additional for flavor or added health benefits -L. acidophilus and/or Bifidobacterium spp.

Procedure
1. Sterilize the cheese pot, spoons, thermometer, bowls, and if using, incubating glasses. Properly dilute sanitizing bleach solution (2 tablespoons bleach to 4 quarts water) and drain equipment thoroughly. Use boiling water technique as an alternative. See Sanitation on page 151.
2. Warm milk in a cheese pot or sauce pan over medium heat to 165°F (73.9°C). Stir gently to prevent scorching on the bottom of the pan.

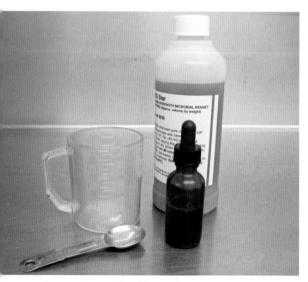
Fermentation–produced Chymosin

3. Turn off the heat, cover the pan, and let sit for 10 minutes.

4. Remove from the heat and immerse in a cold water bath, reducing the temperature to 118°F (47.8°C). Stir or whisk frequently to quicken the cooling process. **Option**: If using gelatin, remove ½ cup of the warm milk and sprinkle gelatin on top. Set aside to soften while the milk cools.

5. Allow the milk to continue cooling to 115°F (46.1°C). Sprinkle the dry yogurt culture and optional skim milk powder over the surface of the milk and let sit for 4 to 5 minutes to rehydrate. If using fresh yogurt culture, remove ½ cup of cooled milk into a small bowl, and whisk in the fresh yogurt. Carefully spoon this mixture back into the pot of cooled milk. Blend the culture (dry or fresh) into the milk by using 15 to 20 down-and-up strokes with a long-handle spoon or skimmer. **Option:** Fresh culture can be added to softening gelatin/milk mixture when milk has cooled. If using dry culture and gelatin, simply spoon the softened gelatin into cooled milk and blend at the same time as the culture. **Option:** If using rennet, add 1 tablespoon of the mixture to the cooled milk and blend with the culture. **With options:** Keep records on the amount of optional ingredients you used. Adjust the recipe for the desired consistency. More of any optional ingredients will make a thicker yogurt.

Cover the pot or pour cultured milk into a clean, warm bowl or glasses and cover for incubation. Let set for 4 to 6 hours or until firm. A longer set time will result in a firmer, more acidic yogurt. Ideal incubation temperature is 110°F (43.3°C) +/-5 Don't let the temperature go below 98°F (36.7°C) or above 125°F (51.7°C).

6. *Notes:* A pre-warmed oven, warming blanket, or heavily wrapped towels will work to keep the yogurt at a proper temperature. Keep the incubating yogurt out of cold drafts. A commercial yogurt maker keeps a steady temperature. Pre-warm the unit while you heat and cool the milk, following manufacturer's instructions. When experimenting with an incubating setup, heat a cup of water to 115°F (46.1°C) and set it next to your incubating bowl/pot. Place your dairy thermometer in the water and monitor now and again to check the temperature of the environment. Make adjustments as necessary.

7. Refrigerate finished yogurt for several hours or overnight before using.

Note: When using fresh yogurt culture, be sure the yogurt has live active cultures with no additives or preservatives. Previous batches of your own yogurt can be used for the culture. Use 10-percent-by-volume for reculturing. Replace with new culture every four to six batches, as the acidity will increase.

Enjoy fresh goat yogurt with fruit, cucumbers, applesauce, honey, maple syrup, and cereal. Make a refreshing and nutritional drink by mixing whey from cheese making with the yogurt. Muddle some fresh mint from your garden for a summer treat!

KEFIR

Kefir is a fermented milk drink that has both acidity and a bit of carbonation. The fermenting organisms are lactic bacteria and yeast. This "champagne of milk" is easy and fun to make, and yes, ripened kefir does contain traces of alcohol. Kefir can be made with either culture or gelatinous, lumpy "grains." I find the grains method most successful. You can buy kefir grains from suppliers, but sharing amongst friends makes for the best-tasting kefir. Due to the multiplying nature of kefir grains, most everyone who makes this fermented dairy product eventually has grains to share. Sandor Ellix Katz, in *The Art of Fermentation*, states, ". . . it [kefir] relies upon a SCOBY, a rubbery mass of bacterial and fungal cells that has evolved an elaborate symbiotic arrangement, sharing nutrients, coordinating reproduction, and co-creating a shared form, which is not microscopic." He goes on to say that biologist Lynn Margulis ". . . explains that kefir grains involve a community of 30 different types of microbes, including common food fermentation favorites, such as Lactobacilli, Leuconostoc, Acetobacter, and Saccharomyces, as well as others more obscure; in fact, . . . fewer than half the microbes involved are known or named." How amazing! Even in this technology and information age, we can easily make and enjoy something that remains a bit mysterious!

Goat milk kefir, as well a being refreshing, is a healthy dairy drink. Kefir is known for boosting the immune system and helping to cleanse and balance the whole body. The beneficial microbial flora are important for intestinal health and digestion. Kefir calms the nervous system and helps people with various disorders, such as bipolar conditions, sleep and actually function.

Kefir is an amazing goat milk product that is so easy to make and enjoy! Make a batch and you, too, will become a kefir keeper. Cheers!

Kefir grains separated from ripened kefir.

Equipment
- 2 1-quart glass jars, see note[1]
- Breathable mesh cover, see note[2]
- Strainer, plastic mesh, see note[3]
- Spoon, wooden, see note[2]

Ingredients
- 3 cups goat milk
- 1 tablespoon kefir grains

Procedure
1. Thoroughly wash and rinse all equipment with clean, non-chlorinated water and allow to dry.

2. Add grains to glass jar.

3. Add milk.

4. Cover with a breathable mesh.

5. Set the jar at room temperature, out of direct sunlight or cold draft, for 24 to 36 hours. Gently stir or swirl to mix milk and grains several times during this period.

6. Allow to ripen to your desired taste and thickness. Cooler ambient temperatures will require a longer ripening period. Use strainer or spoon to separate grains from the milk.

7. Place grains in a new, clean jar, transferring the grains with a small amount of ripened kefir. Add milk to start again.

8. Store the finished kefir in a glass jar with a tight-fitting lid.

This can rest at room temperature for 1 to 2 hours to accumulate carbonation, or you may refrigerate it immediately. Gentler carbonation will form at the refrigerated temperature.

Notes:
- Always give kefir room to expand. This thriving, living food is actively fermenting and needs space to froth and grow.
- I simply use a clean crocheted dish cloth (I have two that I use for only this purpose) held over the rim of the jar with a rubber band.
- Many say kefir should not contact metal at all. Katz agrees that prolonged contact with metal should be avoided due to the acidic nature of kefir. In regards to brief metal contact during the procedure, he has not found that metal destroys the kefir grains. Set up your kefir-making equipment with what is available to you. Affordable equipment that can be easily and thoroughly cleaned is most important.

Chevre is as tasty as it looks—try it with a wide variety of recipes.

Kefir grains will grow and expand. Recommended grain to milk ratio is no more than 10 percent. Besides sharing with friends, you can blend grains into smoothies, feed them to pets, or dehydrate them for future use.

CHÈVRE

Chèvre is a simple cheese—an elegant expression of the complexities of goat milk. Fresh milk of high quality and composition will yield a delicate-tasting cheese with just the right balance of acidity and salt. Taste nuances will alter with the seasons if the goats are eating local plants and browse. Drizzle some honey and serve with an apple or pear. Chunk your fresh chèvre on your favorite salad or pizza. Sunday morning eggs will wake up with a bit of chèvre softened on top. Recipes abound on the Internet and in cookbooks. Pack the fresh goat cheese in a school or work lunch. This simple cheese can be portioned and frozen, ready to put into a lunch cooler before you head out in the morning. The cheese will be thawed and ready to enjoy at noontime.

Equipment

Day 1
- Cheese pot (heavy-bottom stainless steel pot or saucepan)
- Long-handled spoon or skimmer
- Dairy thermometer
- Small bowl or measuring cup
- Measuring spoons

Day 2
- Ladle or stainless steel measuring cup
- Cheesecloth bags or cloth-lined colander for draining
- Whey-catching basin or bucket
- Optional: Cheese molds

Day 3
- Mixing bowl
- Measuring spoons
- Mixing spoon or fork
- Scale

Ingredients
- 8 quarts goat milk, pasteurized or raw
- ¼ teaspoon mesophilic culture
- 1 drop liquid rennet
- Cold, non-chlorinated water for rennet dilution
- Kosher or sea salt
- Optional: herbs, spices, dried fruit or vegetables

Chèvre culture is mesophilic bacteria: generally a mixture of *Lactococcus lactis* ssp. *lactis* and *Lactococcus lactis* ssp. *cremoris*. Early lactation chèvre is enhanced with the addition of *Lactococcus lactis* ssp. *diacetylactis* and/or *Leuconostoc mesenteroides* ssp. *cremoris*. Culture suppliers generally offer several different blends. Experiment and see what works with your goats' milk for your tastes.

Procedure

DAY 1
1. Sterilize today's equipment. Properly dilute sanitizing bleach solution (2 tablespoons bleach to 4 quarts water) and drain equipment thoroughly. Use the boiling water technique as an alternative. See Sanitation on page 151.

2. Warm milk gently to 84°F (28.9°C) over medium heat, stirring to prevent scorching. Remove from heat.
3. Sprinkle culture over the milk and allow it to rehydrate for 4 to 5 minutes.
4. Blend the culture into the milk by using 15 to 20 down-and-up strokes with a long-handle spoon or skimmer.
5. Dilute the rennet in 1 tablespoon of cold water. Add this to the milk, again with 15 to 20 down-and-up strokes. Cover and let set at room temperature, out of drafts, for 20 to 24 hours.

Day 2
1. Sterilize today's equipment. Prepare the cheesecloth or molds. See Sanitation on page 151.
2. Curd surface should be smooth, like a baby's cheek, with no or few cracks or fissures. Carefully press the ladle onto the surface without breaking it to remove the surface whey. Discard what you gather into the whey bucket.
3. Gently ladle curd into the cheesecloth bag, colander, or molds. Ladle horizontally across the surface and work your way to the bottom of the cheese pot, being careful to not break up the curds. Work swiftly and deliberately.

Dilute rennet in cold, chlorine-free water.

4. Hang the bag or place the colander or molds on a draining mat to drain whey for 12 to 24 hours, until the desired texture is attained. The time will also depend on atmospheric conditions and milk composition. Cheese will firm when chilled. Keep records and note changes. Curd placed in molds should be flipped at 6 to 8 hours.

Day 3
1. Sterilize today's equipment. See Sanitation on page 151.
2. Remove drained curd from bags, colander, or molds. Place it in a mixing bowl. Weigh the drained curd, and add 1 percent of the weight in salt (see note below). Mix gently and thoroughly.
3. Form cheese into shape or simply keep it in the bowl.
4. **Optional:** Mix in optional ingredients or press them onto the outside of a formed cheese.
5. Chill cheese in the refrigerator for 6 to 8 hours before eating.

Note: One teaspoon of salt weighs roughly .20 ounce. Example: 2 pounds of cheese = 32 ounces. One percent of 32 is .32 ounces, so use 1½ teaspoons of salt with 2 pounds of chèvre. Record the amount of salt in your cheese log, and adjust as needed. Use only salt with no iodine, such as kosher or sea salt.

 Water-soluble riboflavin (vitamin B2) is responsible for the yellowish-green color of whey.

FETA
Feta is often the cheese that people think of when they hear "goat cheese." Feta is in fact a type of semi-firm, brined cheese that is made with sheep, goat, cow, or a combination of milk. It has been made for centuries in the Mediterranean region by many groups of people, each using area traits to create their local "white" or "pickled" cheese. Traditionally, the regional feta is often made with whatever dairy animal is most popular. Greek feta is

Cheesecloth bags hanging to drain the whey.

Feta dressed with rose petals.

primarily sheep milk, French feta is made from goat and sheep milk, and American feta is generally from cow milk. Feta uses mesophilic cultures, and if desired, a lipase that specifically breaks apart milk fats. The lipase unfolds a complex flavor profile, often described as piquant or sharp. This is the distinctive feta flavor—that puckery salty bite with a dry but smooth texture. Try feta in a watermelon salad, traditional spinach pie, or marinated in olive oil with rose petals or herbs.

Equipment

- Cheese pot (heavy-bottom stainless steel pot or saucepan)
- Long-handled spoon or skimmer
- Dairy thermometer
- Measuring spoons
- Measuring cup
- Long knife or wire for cutting curd
- Cheese molds or draining pans
- Whey-catching basin or bucket
- Food-grade plastic tub, with cover, for brine

Ingredients
- 8 quarts goat milk, pasteurized or raw
- ¼ teaspoon mesophilic culture
- Optional: ⅛ teaspoon kid lipase
- ½ teaspoon rennet
- Cold, non-chlorinated water for rennet dilution and brine
- 1 pound non-iodine salt, for brine

Procedure
1. Sterilize all equipment. See Sanitation on page 151.
2. Gently warm milk to 86°F (30°C) over medium-low heat. Stir to prevent scorching.
3. Remove from heat and sprinkle culture on milk surface to rehydrate for 2 to 4 minutes.
4. Blend culture into milk with 15 to 20 down-and-up strokes. Cover and ripen in a warm place for 45 to 60 minutes. Keep temperature around 86°F (30°C). **Optional:** If using lipase, sprinkle it on the milk surface after 45 minutes and incorporate it into the milk with 15 to 20 down-and-up strokes. Allow the milk to sit for an additional 15 minutes.
5. Dilute rennet with 1 tablespoon of cold water and add it to the milk mixture. Stop the movement of the milk after 10 strokes, or when it is well blended, by holding the spoon steady in the milk and then gently removing it. Cover and let sit for 30 to 45 minutes, allowing the curd to form.
6. Test the curd for a clean break, and then cut the curd into ½-inch cubes. Let curds rest for 5 minutes, releasing whey and firming.
7. Keeping a steady temperature of 86°F (30°C), gently stir the curds for 20 to 30 minutes. The longer time will result in a firmer cheese. *Note:* To maintain a steady temperature, fill a sink or large tub with water warmed to 86°F (30°C). Place your cheese pot in this water bath, and adjust the water temperature as needed.
8. Allow the curds to settle to the bottom of the cheese pot for 5 minutes, and then pour or ladle off the surface whey. Ladle the soft curds into pans or molds, filling them to the desired thickness. Allow to drain for 6 to 8 hours, flipping the cheese 3 to 4 times in the first 2 hours to allow for even drainage. Weight pressing is not necessary for feta. You can stack the pans on each other and rotate weight as the whey drains.
9. Transfer cheese to a brine now, or if a drier, firmer cheese is desired, allow it to drain, without flipping, for an additional 12 hours.
10. Make a light to medium brine. Recommended ratio is 16 oz. (1 pound) kosher or sea (non-iodized) salt to 4 quarts water. See notes below.
11. Remove cheese from the mold/pans and place it in the brine solution at 55 to 60°F (12.8 to 15.6°C) for 3 to 4 days. Flip the cheese in the brine on the second day to encourage even salt penetration.
12. Remove cheese from the brine and refrigerate.

Notes:
[1] Place a clean finger or spoon into the ripening curd and lift. The curd should have a definitive separation.

If storing in brine for future use, the brine must be properly maintained to keep the correct percentage of salinity. See more information in *The Cheesemaker's Manual,* by Margaret P. Morris, or in other cheese-making resources.

CROTTIN
This is a classic French goat cheese that is a small gem of smooth paste inside a white mold crust. The joy of this cheese is its ever-changing characteristics. Cheeses can be eaten early after making, with subtle mold growth and full moisture. With some ripening, the paste firms and a more definitive skin develops. Continue ripening, and the cheese, with proper conditions, continues to dry and harden, becoming sweet and dense. The skin dries and matures, with yet again a different texture to the inside paste. Enjoy this cheese in its many stages. An 8-quart batch should yield around twenty 2-ounce cheeses. This will be enough to eat and share with family and friends.

Equipment
- Cheese pot (heavy bottom stainless steel pot or saucepan)
- Long-handled spoon or skimmer
- Dairy thermometer

- Measuring spoons
- Measuring cup
- Ladle
- 20 crottin molds
- Whey-catching basin or bucket
- Draining container
- Ripening container
- Cheese matting

Ingredients
- 8 quarts milk, pasteurized or raw
- ¼ teaspoon mesophilic culture
- ⅛ teaspoon *Penicillium candidum* mold powder
- 1/16 teaspoon *Geotrichum candidum* 15 mold powder
- 8 drops liquid rennet
- Cold, non-chlorinated water for rennet dilution
- Optional: ⅛ teaspoon calcium chloride
- Kosher or sea salt

Procedure
1. Sterilize all equipment. See Sanitation on page 151. Prepare an area for mold filling and draining.
2. Gently warm milk to 72°F (22.2°C) over medium low heat. Stir to prevent scorching.
3. Remove from heat and sprinkle culture and both mold powders on milk surface to rehydrate for 2 to 4 minutes.
4. Blend culture and molds into milk with 15 to 20 down-and-up strokes.
5. *Optional:* Dilute calcium chloride in ¼ cup cool water. Add to milk.
6. Dilute rennet with 1 tablespoon of cold water and add to milk mixture with similar down-and-up strokes.
7. Cover the pot and let the contents ripen at room temperature for 18 to 20 hours. A firm curd should form with a small amount of surface whey.
8. Remove surface whey by carefully pressing the ladle onto the surface and gathering whey without breaking the surface. Discard into whey bucket.
9. Gently ladle curd into prearranged molds. Ladle horizontally across the curd surface and work your way to the bottom of the cheese pot, placing thin "slices" of curd

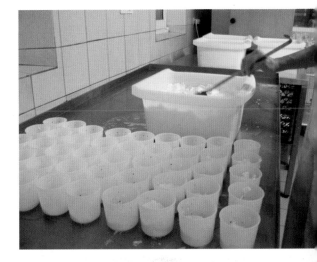

Crottin

into each mold. Layer one mold completely before moving onto the next. Be careful to not break up the curds. Work swiftly and deliberately. Return to fill molds that are draining, being careful to evenly distribute the soft curds. Strive to create cheeses of similar size. Avoid adding additional molds—this will result in small, irregular cheeses.

10. Let cheeses drain at room temperature for 24 hours, flipping them a couple of times when firm. This allows for better drainage of the whey. Cover the cheeses lightly to protect them from dust and drying. A light cover also helps to maintain steady temperature.

11. When the cheeses are finished draining whey, generally 24 to 36 hours after ladling, carefully remove them from the molds. Sprinkle tops and bottoms each with ¼ teaspoon salt. Develop a rhythm with a salt shaker to make this task go more quickly and distribute the proper amount of salt evenly. This takes some practice, but is well worth the mistakes and the time.

12. Return salted cheeses to a clean drainage mat and allow to drain again at room temperature for 24 hours. Flip the cheese once during this time to permit whey accumulated on the top of the cheese to drain as well.

13. Transfer cheeses to a ripening area and let them ripen at 50 to 54°F (10 to 12.2°C) and 88 to 90 percent humidity for 2 weeks. The white mold should begin to show by day 5 to 7 in ripening conditions. The cheeses need to be flipped every day at this point to allow for even growth of the white mold skin.

14. When completely covered in mold skin, the cheeses can be wrapped in a breathable paper and moved to a refrigerator for storage or to be eaten.

15. Continue to ripen some cheeses for an additional week.

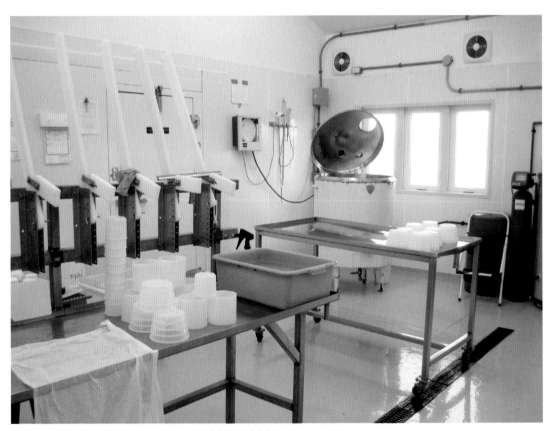

Clean supplies and working environment are crucial to good dairy.

Top and above: The breadth of delicious dairy products from goats is as endless as the varieties of goats themselves.

Chapter 12

The World of Dairy Goats

Dairy goats have provided food and companionship to humankind for centuries. They are curious, adaptable, and friendly animals. In addition to adding wholesome food to our tables, they help bring us together. My experience with goats has opened an interesting new world: the world of dairy goats and their keepers.

And what a vast world it is! A countryside farmer in China who uses goat pellets to fertilize his prized white peach trees; a 500-head goat farm in Minnesota meticulously managed by dedicated family and staff; a few goats on a half-acre lot behind a historic home in a Boston suburb; goats being milked in the Provence region of France for making the traditional chestnut-leaf-wrapped, eau de vie-soaked Banon cheese; settlement goats brought initially to Iceland centuries ago now being stewarded and thoughtfully bred to keep nearly extinct genes alive and productive; children working and learning with parents and friends to improve and show their goats; dairy goats clambering up the rocky inclines of the Canary Islands; loyal people enhancing lives and community through improvements of dairy goats in Grenada; and many small productive herds keeping their keepers happily smiling and healthfully employed in many areas of the world.

The thread holding this tale together is the dairy goat. Goats and their keepers speak a universal language. While many breeds have adapted and flourished worldwide, dairy goats share a common bond; that inquisitive nature, that while seeming common and unassuming, is really quite complex and intelligent.

Dairy goats possess qualities that make them ideal food-producing animals for many areas around the world. Their efficient production helps provide a family with nutritious food, as well as income to help balance living expenses. Dairy goats play a beneficial role in the world environmental concerns we currently face—they are productive while requiring less water than other farm animals, their carbon "hoofprint" is small, and goats can turn the invasive plant species that plague many areas into useful milk and meat. Given some gentle care, dairy goats nurture their keepers with a calming, thoughtful nature. The simple relaxing routine of milking dairy goats settles and balances us. Sipping their sweet milk restores the precious feeling of just being. Yes, they have spirit and spunk, but spend a bit a time working with goats and looking into their eyes, and you will quickly learn important lessons of life. You will develop GOAT-titude!

Opposite and left: Dairy goats and goat keepers speak a universal language. That language is often motivated by food.

Goats may insist on more food even when they're not hungry. Be wary!

Goats will greet just about every guest.

Remember to clean and sterilize all interior dairy-producing tools and surfaces.

Goats are as curious about you as you are about them. Watch out!

At state and county fairs, goats can offer pleasure and comfort to young and old alike.

Poised and pointed ears reveal an inquisitive and curious nature. *Jay Iversen*

State and county fairs are a fine opportunity to showcase your finest goats and their dairy products.

Mmm—goat cheese!

Regular pen maintenance is important for your goats' health and happiness.

It's a goat apocalypse!

Dairy goats enjoy exercise and late-season foraging.

Many breeds of dogs are genetically wired to keep your herd safe.

Goats respond to speech and touch—be nurturing and kind for a positive experience.

A wise and observant goat keeps her eyes on the surroundings. *Jay Iversen*

Newborn kids can be sticky affairs—wear your galoshes or an old pair of tennis shoes! *Danielle Mulcahy*

A common sight at the goat farm. Bottoms up! *Danielle Mulcahy*

The curious doeling greets visitors with a gleam in her eyes.

Kits

Devote a labeled plastic tote to each kit. Copy the list and tape it inside lid, writing additional items as needed. Pull out your birthing kit several weeks before birthing to resupply. These items can be purchased from livestock/goat supply catalogs and online stores. Many items can be found in a local agricultural store or pharmacy. Some are only sold in large quantities or are much cheaper that way. Cooperative buying is a good idea.

Birthing Kit
- Clean, large towels (oldies but good still)
- OB lube
- Long-sleeved gloves (AI shoulder-length gloves)
- Gloves (nitrile or latex)
- 7 percent iodine and cup for navel dipping
- Clean scissors
- Flashlight
- Thermometer
- Black strap molasses
- Propylene glycol
- Nutri-drench or similar product
- Paper, pencil/pens
- Old woolen or polyfill vest (small for kids, large for does)
- Phone numbers for vet and neighbors

Kid-Feeding Kit
- Nipples and bottles (plastic soda, beer, 2-quart. calf)*
- Colostrum cubes frozen from previous year
- Instructions and vessels for heat-treating (CAE prevention)
- Weak Kid drenching syringe and tube (catheter)
- Milk Replacer (good quality goat if you choose to use this)
- Liquid measuring cup
- Metal whisk for stirring
- Lambar (include nipples, tubes, and thin brush for cleaning)
- Towels

* Even if you plan to dam-raise your kids, have a couple on hand for ones that need hand feeding.

First Aid Kit and Medications

- Thermometer(s)*
- Drench syringe
- Syringes (variety of sizes - 3cc, 12cc, 35cc, 60cc)
- Needles (20 gauge, 1 inch)
- Scalpel and blades
- Small gauze pads or clean cloth strips
- Vet wrap or bandaging
- Clean towels (small and large)
- Electrical tape
- Betadine wash (not scrub) or iodine for washing wounds
- Rubbing alcohol
- Hydrogen peroxide
- Granulated garlic (not garlic salt!), blood stop, or antiseptic powder**
- Clean, sharp surgical scissors
- Copies of Goat Vitals and Anatomy from this book
- Aspirin
- Electrolytes
- Saline solution for eyes
- Bloat Release
- Pen G Procaine (keep refrigerated)
- Oxytetracycline
- Vitamin B Complex
- Activated charcoal
- Herbal/homeopathic remedies
- Thiamine (B1 [prescription required])
- Bo-Se (prescription required)
- Epinephrine
- Tetanus anti-toxin
- Phone numbers for vet, poison hot line and neighbors

* I prefer the old-fashioned glass thermometers (tie a string with clip on end to attach to tail). Digital thermometers are easy to use. I keep and use both.

** Any of these will help stop bleeding for hoof trimming injuries.

Barn Tool Kit*

- Hammer
- Set of screwdrivers
- Adjustable wrench
- Vise grips
- Wire cutters
- 3-in-1 oil
- WD-40
- Variety of screws and nails
- Small roll of wire
- Electric fence supplies
- Flashlight and batteries
- Crowbar
- Bolt cutters
- Saw

*Color code these with electrical tape and "Return to Goat Barn."

Resources

Books

200 Easy Homemade Cheese Recipes
Debra Amrein-Boyes

A Guide to Starting a Commercial Goat Dairy
Carol Delaney
Free as a download or order from
www.uvm.edu/sustainableagriculture

The Art of Fermentation
Sandor Ellix Katz

Cheese and Fermented Milk Foods, 2nd edition
Frank Kosikowski

The Cheesemaker's Manual, 2nd edition
Margaret P. Morris

Cheesemaking Practice, 3rd edition
R. Scott, R.K. Robinson, and R.A. Wilbey

Extension Goat Handbook
George Haenlein and Donald Ace (USDA)

The Fabrication of Farmstead Goat Cheese
Jean Claude LeJaouen

Goat Husbandry
David Mackenzie

Goat Medicine, 2nd edition
Mary Smith and David Sherman

Herbal Handbook for Farm and Stable
Juliette de Baïracli Levy

How to Build Animal Housing
Carol Ekarius
Natural Goat Care
Pat Coleby

Nutrient Requirements of Small Ruminants
National Research Council

Raising Goats for Milk and Meat
Rosalee Sinn and Paul Rudenberg

Websites and Periodicals

American Consortium for Small Ruminants Parasite Control
www.acsrpc.org

American Dairy Goat Association
www.adga.org

American Goat Society
www.americangoatsociety.com

Appropriate Technology Transfer for Rural Areas
www.ATTRA.NCAT.org

BioTracking, LLC (pregnancy confirmation tests for goats)
www.biotracking.com

The Biology of the Goat
www.goatbiology.com

Caprine Supply Company
www.caprinesupply.com

Cornell University Diagnostic Laboratory
ahdc.vet.cornell.edu

Cornell University Poisonous Plants Database
www.ansci.cornell.edu/plants

Culture (quarterly cheese magazine)
www.culturecheesemag.com

Dairy Goat Journal (monthly periodical)
www.dairygoatjournal.com

Dairy Herd Improvement Association
www.dhi.org

Dairy Practices Council
www.DAIRYPC.org

E (Kika) de la Garza American Institute for Goat Research
www.luresext.edu/goats/

Get Culture cheese-making supplies
www.getculture.com

Goat Dairy Library
Goatdairylibrary.org

Hoegger Supply Company
www.hoeggerfarmyard.com

International Goat Association
www.iga-goatworld.com

Jeffers Livestock Supply Company
www.jeffers.com

Lancaster Agriculture Products
www.lancasterag.com

Maryland Small Ruminant Page
www.sheepandgoat.com

National Institute for Animal Agriculture
www.eradicatescrapie.org

New England Cheesemaking Supply Company
www.cheesemaking.com

Premier1 Supplies
www.premier1supplies.com

United Caprine News (monthly periodical)
www.unitedcaprinenews.com

University of Vermont Center for Sustainable Agriculture
www.uvm.edu/sustainableagriculture

Washington State University Diagnostic Laboratory
waddl.vetmed.wsu.edu

Dairy Animal Record Sheets

Dairy Doe Record Sheet

Breed   Reg. #   Born

Name   Tattoo #   Disbudded

Color and Description

Sire   Dam

Reg. #   Reg. #

Sire   Dam   Sire   Dam

Reg. #   Reg. #   Reg. #   Reg. #

Origin - Farm Raised

Purchased From   Date   Cost

Date Died

Date Sold   To   Cost

LACTATION

Birthing Age	Total Days	Pounds Milk	Butterfat	Protein

KIDDING RECORD

Bred	Sire used	Due	Kidded	#, sex kids born	Notes

HEALTH

Date	Condition	Treatment

Date	FAMACHA score	Body Condition Score	Notes

BREEDINGS

Date	Doe	Notes

Glossary

Abomasum - This is the true stomach of the gastrointestinal system. Lining cells secrete enzymes and acids that aid in breaking down digested feed.

Alpine - A breed of medium- to large-sized dairy goat with upright ears. There is a variety of acceptable hair coat colors and pattern combinations. They average milk production of 3-4 quarts (liters) per day, 3.5% milk fat, and 2.9% protein.

Artificial Insemination (A.I.) - Manual breeding of a goat by placing sperm that was previously collected from a buck into a doe's uterus by means of sterile equipment. The sperm is usually stored frozen in labeled straws and can be stored as such for long periods of time.

Bloat - A buildup of naturally occurring fermentation gas within the rumen. Bloat left untreated can cause death.

Browse - Vegetation, such as woody stems, twigs, vines, and other high-growing plants, that is preferred feed for goats.

Buck - A male goat.

Buckling - A young male goat, under one year of age.

CAE (Caprine Arthritis Encephalitis) - CAE is a viral disease of goats mainly spread through feeding kids colostrum and milk from infected animals. Symptoms in older animals include arthritis, pneumonia, and cystic (hard) udders. Young kids generally show signs of nervous system difficulties, such as leg paralysis and encephalitis.

Casein - A milk protein. The casein protein is a complex matrix structure of protein, calcium, and phosphorous.

Castration - Neutering of males by removing the testicles or separating the connection of spermatic cords from the testicles.

Chèvre - Fresh, unaged goat cheese.

Colostrum - The first milk produced after a doe gives birth. This fluid is high in passive immune antibodies that are important for helping newborn kids fight infections until their own immune systems are active.

Curd - The solid fraction of milk after coagulation alters the milk proteins.

Dairy Herd Improvement Association (DHIA) - an organization that helps dairy farmers, including those milking dairy goats, improve their farm business, and their animals' production and health. Monthly milk samples are submitted and analyzed, and the results are reported back to the owners for use in making management decisions.

Dental pad - Firm soft tissue at the front of the upper jaw that helps the incisors (lower front teeth) tear forage.

Doe - A female goat.

Doeling - A young female goat, under one year of age.

Disbud – The act of destroying the horn bud cells by applying heat so the horns will not grow.

Estrus (Heat) - The period of ovulation, generally lasting 24 - 48 hours, when the female's ovary releases eggs, which are capable of being fertilized by male sperm.

Estrous Cycle - The reproductive period between two estrus dates. The cycle runs an average of 21 days and includes four main phases.

FAMACHA - A low-cost system for quickly assessing a goat for anemia, which could possibly indicate a high level of parasitism. The color of a goat's inner eye membrane is compared to a reference card, ranging from white (anemic) to red (healthy).

Gestation (pregnancy) - The time required for a fetus to develop from fertilization to birth. Average time is 150 days for dairy goats.

Heart girth - Measurement around the fore barrel of the goat, behind the front legs and shoulders.

Kid - A young goat, generally under one year of age.

Lactose - Naturally occurring sugar found in milk, made up of glucose and galactose.

LaMancha - A medium-sized breed of dairy goat with short or basically no external ear. Hair coat is any or all colors, solid or patterned. They average 3 quarts (liters) per day, 3.7% milk fat, and 3.1% protein.

Lambar - A bucket with many attached nipples, allowing multiple kids to feed at the same time.

Mastitis - Inflammation of the mammary gland (udder), generally caused by pathogenic bacteria or yeast.

Mesophilic bacteria - Bacteria thriving in a mild temperature, with an ideal range of 77–86 degrees F.

Nigerian Dwarf - A small-sized breed of dairy goat with upright ears. Hair coat is any or all colors, solid or patterned. They average .5 - 1 quart (liter) per day, 6.3% milk fat, and 4.3% protein.

Nubian - A medium- to large-sized breed of dairy goat with pendulous ears. Hair coat is any or all colors, solid or patterned. They average 2–3 quarts (liters) per day, 4.5% milk fat, and 3.7% protein.

Oberhasli - A breed of medium-sized dairy goat with upright ears. Hair coat is bay (reddish-brown) with black markings on face, back, belly, lower legs, and tail. They average milk production of 3 quarts (liters) per day, 3.6% milk fat, and 2.9% protein.

Omasum - The smallest compartment of the stomach system. The multi-fold lining absorbs water and minerals from the digested feed.

Parturition (kidding) - Giving birth.

Polled - Naturally hornless goat. Use caution when breeding two polled goats. The resulting offspring may be incapable of reproducing.

Rennet - Enzymes produced naturally in the stomach of young goats that coagulates the milk proteins, turning liquid milk into a more solidified mass.

Reticulum - A chamber in the goat's four-compartment stomach system. This compartment has a honeycomb-patterned lining. The reticulum helps mix and move feed through the gastrointestinal system.

Rumen - The largest chamber of the goat's four-compartment stomach system. Fermentation and break down of plant fibers occurs in this chamber.

Rut - The time of year, usually with the beginning of increased darkness, when male goats experience a surge of hormones and are primed for breeding does. Signs of rut include increased release of musk from their scent glands, urinating on their front legs and head, increased aggression, and a possible decrease in appetite.

Saanen - A breed of medium- to large-sized dairy goat with upright ears. Hair coat is solid white or cream with occasional black spots on nose, ears, and udder. A color variant, known as Sable, is recently accepted by goat associations. They average milk production of 3–4 quarts (liters) per day, 3.3% milk fat, and 2.8% protein.

SCOBY - A symbiotic colony of bacteria and yeast used to make kefir, a fermented dairy product.

Somatic Cell Count (SCC) - A milk quality test that indicates overall health of the udder. SCC counts leukocytes or white blood cells in the milk with high counts indicating a mammary infection.

Tattoo - Permanent identification made by piercing the ear or tail web with specific needled symbols and smearing with tattoo ink.

Thermophilic bacteria - Bacteria thriving in a higher temperature, with an ideal range of 95–105 degrees F.

Toggenburg - A breed of medium-sized dairy goat with upright ears. Hair coat is light brown with white markings on face, ears, muzzle, legs, and tail. They average milk production of 3 quarts (liters) per day, 3.2% milk fat, and 2.7% protein.

Udder - The milk producing gland of the goat. This gland is split into two compartments, each completely walled off from the other, and milk is secreted through a teat located at the bottom of each portion.

Urinary Calculi - A concentration of salt crystals in the urinary tract that cause a painful and possibly lethal blocking. This condition is generally seen in castrated males.

Wattles - Small skin pendant hanging downward, generally on the neck or by ears. Wattles serve no function and are not found on all goats.

Wether - A castrated male goat.

Whey - The liquid fraction of milk after coagulation alters the milk proteins.

Withers - Ridge on the back between shoulder blades.

Yearling - A young goat, between one and two years of age.

Index

About the Author

Ann Starbard owns and operates Crystal Brook Farm, a farmstead goat cheese dairy in central Massachusetts. Her 70 four-legged co-workers bring joy and challenge to each day, fortifying her GOAT-titude. Teaching people to provide nutritional food for themselves and others is a part of Ann's life- long mission. She first learned about dairy goats working for Heifer International, gathering and shipping goats worldwide to help alleviate hunger and poverty. Sharing the benefits of dairy goat milk and wholesome milk products and realizing the real food value from these adaptable farm animals is important to Ann.

Working with dairy goats to improve natural resources is also a priority. Ann shares her expertise and knowledge while serving on local agricultural and land conservation boards, including 25 years as supervisor for the Worcester County Conservation District. She has traveled to France, Grenada, China, Iceland, the Canary Islands, and throughout many areas of the United States studying various models of raising dairy goats, making cheese, and employing goats to enhance the local environment and economy.

Ann holds a B.S. with Honors in Animal Bioscience from The Pennsylvania State University. She leads workshops and training seminars on dairy goat raising, farmstead cheese making, and developing business and marketing plans for small-scale agricultural ventures.

Ann has a deep connection to agriculture, feeling fortunate to have grown up in a multi-generational active farm family in south central Pennsylvania. She carries on her family's tradition of love for the land and sense of place, as well as care and compassion for the earth's animals, plants, and people.

Learn more at www.GOAT-titude.com.

Acknowledgments

I am grateful to the many people who have shared their knowledge, time, and support with me to create this book. My initial goat-raising mentors—Rosalee Sinn, David Sherman, and Mary and Dan Fox—set a foundation that I cherish more than they know. I am thankful for their continued gifts of friendship and expertise. We all have to start somewhere.

Thank you to the people I visited and interviewed, including Joël and Brigitte Corbon, Carol Delaney, Susan Phinney, Tricia Smith, Erin DeCoste, Jenn Janowicz, Jóhanna Bergmann, Nina Sharp, Judy Schnell, Jennifer Poirier, The Maefsky family, and so many more. Visiting other farms, goat keepers, and cheese makers brings invaluable insight to this amazing world of agriculture and dairy goats. Thank you to Dr. Caitlin Vincent Eaton and the Buck Hill Veterinary team—you are literally life savers. A super-duper thanks to Christine Curry and The Goat Dairy crew, St. Patrick's, Grenada. Three cheers for walking the walk against many odds! Tomorrow is here.

A special thanks to Yvonne Zweede-Tucker for not only finding me a wonderful guardian dog, but also opening this big barn door. Thanks to Danielle Mulcahy and Kate Wollensak Freeborn for sharing your spirited art, photography, and design talents. You are amazing! Thanks also to project manager Jordan Wiklund for jumping in at halftime and guiding me steadily along.

A heartfelt thanks to my family for all the guidance, support, laughter, and love shared during oh so many hours milking and caring for animals. Dad and Mom, you taught us many valuable life lessons along the way. Thank you, Eric, for your help while I was writing. Challenges help us grow.